# Student Solutions Manual

# Introductory Chemistry: A Foundation

### NINTH EDITION

**Steven S. Zumdahl**
University of Illinois at Urbana-Champaign

**Donald J. DeCoste**
University of Illinois at Urbana-Champaign

Prepared by

**Dr. Gretchen M. Adams**
University of Illinois at Urbana-Champaign

**James F. Hall**
University of Massachusetts Lowell

Australia • Brazil • Mexico • Singapore • United Kingdom • United States

© 2019 Cengage Learning

WCN: 01-100-101

ALL RIGHTS RESERVED. No part of this work covered by the copyright herein may be reproduced, transmitted, stored, or used in any form or by any means graphic, electronic, or mechanical, including but not limited to photocopying, recording, scanning, digitizing, taping, Web distribution, information networks, or information storage and retrieval systems, except as permitted under Section 107 or 108 of the 1976 United States Copyright Act, without the prior written permission of the publisher.

For product information and technology assistance, contact us at
**Cengage Learning Customer & Sales Support,
1-800-354-9706**.

For permission to use material from this text or product, submit all requests online at **www.cengage.com/permissions**
Further permissions questions can be emailed to **permissionrequest@cengage.com**.

ISBN: 978-1-337-39947-0

**Cengage Learning**
20 Channel Center Street
Boston, MA 02210
USA

Cengage Learning is a leading provider of customized learning solutions with office locations around the globe, including Singapore, the United Kingdom, Australia, Mexico, Brazil, and Japan. Locate your local office at: **www.cengage.com/global**.

Cengage Learning products are represented in Canada by Nelson Education, Ltd.

To learn more about Cengage Learning Solutions, visit **www.cengage.com**.

Purchase any of our products at your local college store or at our preferred online store **www.cengagebrain.com**.

Printed in the United States of America
Print Number: 02    Print Year: 2018

# Contents

1. Chemistry: An Introduction............................................................................1
2. Measurements and Calculations..................................................................2
3. Matter.............................................................................................................17
   Cumulative Review: Chapters 1, 2, and 3...................................................20
4. Chemical Foundations: Elements, Atoms, and Ions................................23
5. Nomenclature................................................................................................29
   Cumulative Review: Chapters 4 and 5........................................................37
6. Chemical Reactions: An Introduction........................................................41
7. Reactions in Aqueous Solution...................................................................44
   Cumulative Review: Chapters 6 and 7........................................................52
8. Chemical Composition.................................................................................56
9. Chemical Quantities.....................................................................................78
   Cumulative Review: Chapters 8 and 9........................................................99
10. Energy..........................................................................................................103
11. Modern Atomic Theory..............................................................................109
12. Chemical Bonding......................................................................................116
    Cumulative Review: Chapters 10, 11, and 12..........................................126
13. Gases............................................................................................................133
14. Liquids and Solids......................................................................................152
15. Solutions......................................................................................................159
    Cumulative Review: Chapters 13, 14, and 15..........................................179
16. Acids and Bases..........................................................................................186
17. Equilibrium.................................................................................................195
    Cumulative Review: Chapters 16 and 17.................................................204
18. Oxidation–Reduction Reactions and Electrochemistry........................209
19. Radioactivity and Nuclear Energy...........................................................218
20. Organic Chemistry.....................................................................................223
21. Biochemistry...............................................................................................234

# Contents

1. Chemistry: An Introduction .................................................................. 1
2. Measurements and Calculations ............................................................ 2
3. Matter ................................................................................................... 11
   Cumulative Review: Chapters 1, 2, and 3 ............................................. 20
4. Chemical Foundations: Elements, Atoms, and Ions ............................. 22
5. Nomenclature ...................................................................................... 30
   Cumulative Review: Chapters 4 and 5 .................................................. 37
6. Chemical Reactions: An Introduction ................................................... 40
7. Reactions in Aqueous Solution ............................................................ 44
   Cumulative Review: Chapters 6 and 7 .................................................. 52
8. Chemical Composition ......................................................................... 56
9. Chemical Quantities ............................................................................ 72
   Cumulative Review: Chapters 8 and 9 .................................................. 99
10. Energy ................................................................................................. 102
11. Modern Atomic Theory ........................................................................ 106
12. Chemical Bonding ................................................................................ 110
    Cumulative Review: Chapters 10, 11, and 12 ...................................... 126
13. Gases .................................................................................................. 133
14. Liquids and Solids ............................................................................... 152
15. Solutions ............................................................................................. 154
    Cumulative Review: Chapters 13, 14, and 15 ...................................... 175
16. Acids and Bases .................................................................................. 180
17. Equilibrium .......................................................................................... 196
    Cumulative Review: Chapters 16 and 17 ............................................. 204
18. Oxidation-Reduction Reactions and Electrochemistry ........................ 209
19. Radioactivity and Nuclear Energy ....................................................... 218
20. Organic Chemistry ............................................................................... 223
21. Biochemistry ....................................................................................... 234

# Preface

This guide contains the even-numbered solutions for the end-of-chapter problems in the ninth editions of *Introductory Chemistry*, *Introductory Chemistry: A Foundation* by Steven S. Zumdahl and Donald J. DeCoste. New problems and questions have been prepared for the new editions of the text, which we hope will be of even greater help to students in gaining an understanding of the fundamental principles of chemistry.

We have tried to give the most detailed solutions possible to all the problems, even though some problems give repeat drill practice on the same subject. Our chief attempt at brevity is to give molar masses for compounds without showing the calculation (after the subject of molar mass itself has been discussed). We have also made a conscious effort in this guide to solve each problem in the manner discussed in the textbook. The instructor, of course, may wish to discuss alternative methods of solution with his or her students.

One topic that causes many students concern is the matter of significant figures and the determination of the number of digits to which a solution to a problem should be reported. To avoid truncation errors in the solutions contained in this guide, the solutions typically report intermediate answers to one more digit than appropriate for the final answer. The final answer to each problem is then given to the correct number of significant figures based on the data provided in the problem.

We wish you the best of luck and success in your study of chemistry!

Dr. Gretchen M. Adams

University of Illinois at Urbana-Champaign

and

James F. Hall

University of Massachusetts Lowell

# CHAPTER 1

# Chemistry: An Introduction

2. The answer will depend on student examples.

4. Answer depends on student responses/examples.

6. This answer depends on your own experience, but consider the following examples: oven cleaner (the label says it contains sodium hydroxide; it converts the burned-on grease in the oven to a soapy material that washes away); drain cleaner (the label says it contains sodium hydroxide; it dissolves the clog of hair in the drain); stomach antacid (the label says it contains calcium carbonate; it makes me belch and makes my stomach feel better); hydrogen peroxide (the label says it is a 3% solution of hydrogen peroxide; when applied to a wound, it bubbles); depilatory cream (the label says it contains sodium hydroxide; it removes unwanted hair from skin).

8. The scientist must recognize the problem and state it clearly, propose possible solutions or explanations, and then decide through experimentation which solution or explanation is best.

10. Answer depends on student response. A quantitative observation must include a number. For example "There are two windows in this room" represents a quantitative observation, but "The walls of this room are yellow" is a qualitative observation.

12. False. Theories can be refined and changed because they are interpretations. They represent possible explanations of why nature behaves in a particular way. Theories are refined by performing experiments and making new observations, not by proving the existing observations as false (which is something that can be witnessed and recorded).

14. Scientists are human, too. When a scientist formulates a hypothesis, he or she wants it to be proven correct. In academic research, for example, scientists want to be able to publish papers on their work to gain renown and acceptance from their colleagues. In industrial situations, the financial success of the individual and of the company as a whole may be at stake. Politically, scientists may be under pressure from the government to "beat the other guy."

16. Chemistry is not merely a list of observations, definitions, and properties. Chemistry is the study of very real interactions among different samples of matter, whether within a living cell, or in a chemical factory. When we study chemistry, at least in the beginning, we try to be as general and as nonspecific as possible, so that the *basic principles* learned can be applied to many situations. In a beginning chemistry course, we learn to interpret and solve a basic set of very simple problems in the hope that the method of solving these simple problems can be extended to more complex real life situations later on. The actual solution to a problem, at this point, is not as important as learning how to recognize and interpret the problem, and how to propose reasonable, experimentally testable hypotheses.

18. A good student will: learn the background and fundamentals of the subject from their classes and textbook; will develop the ability to recognize and solve problems and to extend what was learned in the classroom to "real" situations; will learn to make careful observations; and will be able to communicate effectively. While some academic subjects may emphasize use of one or more of these skills, Chemistry makes extensive use of all of them.

# CHAPTER 2

# Measurements and Calculations

2. "Scientific notation" means we have to put the decimal point after the first significant figure, and then express the order of magnitude of the number as a power of ten. So we want to put the decimal point after the first 2:

   $2{,}421 \rightarrow 2.421 \times 10^{\text{to some power}}$

   To be able to move the decimal point three places to the left in going from 2,421 to 2.421, means I will need a power of $10^3$ after the number, where the exponent 3 shows that I moved the decimal point 3 places to the left.

   $2{,}421 \rightarrow 2.421 \times 10^{\text{to some power}} = 2.421 \times 10^3$

4. a. $10^7$
   b. $10^{-1}$
   c. $10^{-5}$
   d. $10^{12}$

6. a. negative
   b. zero
   c. negative
   d. positive

8. a. The decimal point must be moved three spaces to the right: 2,789
   b. The decimal point must be moved three spaces to the left: 0.002789
   c. The decimal point must be moved seven spaces to the right: 93,000,000
   d. The decimal point must be moved one space to the right: 42.89
   e. The decimal point must be moved 4 spaces to the right: 99,990
   f. The decimal point must be moved 5 spaces to the left: 0.00009999

10. a. three spaces to the left
    b. one space to the left
    c. five spaces to the right
    d. one space to the left
    e. two spaces to the right
    f. two spaces to the left

Chapter 2: Measurements and Calculations

12. a. The decimal point must be moved 3 places to the right: 6,244
    b. The decimal point must be moved 2 spaces to the left: 0.09117
    c. The decimal point must be moved 1 space to the right: 82.99
    d. The decimal point must be moved 4 spaces to the left: 0.0001771
    e. The decimal point must be moved 2 spaces to the right: 545.1
    f. The decimal point must be moved 5 spaces to the left: 0.00002934

14. a. $1/0.00032 = 3.1 \times 10^3$
    b. $10^3/10^{-3} = 1 \times 10^6$
    c. $10^3/10^3 = 1$ ($1 \times 10^0$); any number divided by itself is unity.
    d. $1/55,000 = 1.8 \times 10^{-5}$
    e. $(10^5)(10^4)(10^{-4})/10^{-2} = 1 \times 10^7$
    f. $43.2/(4.32 \times 10^{-5}) = \dfrac{4.32 \times 10^1}{4.32 \times 10^{-5}} = 1.00 \times 10^6$
    g. $(4.32 \times 10^{-5})/432 = \dfrac{4.32 \times 10^{-5}}{4.32 \times 10^2} = 1.00 \times 10^{-7}$
    h. $1/(10^5)(10^{-6}) = 1/(10^{-1}) = 1 \times 10^1$

16. a. kilo
    b. milli
    c. nano
    d. mega
    e. deci
    f. micro

18. Since a pound is 453.6 grams, the 125-g can will be slightly more than ¼ pound.

20. Since 1 inch = 2.54 cm, the nail is approximately an inch long.

22. $100. \text{ mi} \times \dfrac{1.6093 \text{ km}}{1 \text{ mi}} = 161 \text{ km}$

24. 1.62 m is approximately 5 ft, 4 in. The woman is slightly taller.

26. a. inch
    b. yard
    c. mile

Chapter 2:    Measurements and Calculations

28. b (the other units would give very small numbers for the length)

30. d. 158.5 – 158.7 mL. A measurement always has some degree of uncertainty. Because the last number (the 6 in 158.6 mL) is based on a visual estimate, it may be different when another person makes the same measurement. The first three digits in the measurement (158) are certain numbers of the measurement. However the fourth digit (6) is estimated and can vary; it is called an uncertain number. When one is making a measurement, the custom is to record all of the certain numbers plus the first uncertain number.

32. The scale of the ruler shown is only marked to the nearest *tenth* of a centimeter; writing 2.850 would imply that the scale was marked to the nearest *hundredth* of a centimeter (and that the zero in the thousandths place had been estimated).

34. 
   a. three
   b. unlimited number (definition)
   c. five
   d. two
   e. two

36. It is better to round off only the final answer, and to carry through extra digits in intermediate calculations. If there are enough steps to the calculation, rounding off in each step may lead to a cumulative error in the final answer.

38. 
   a. $1.6 \times 10^6$
   b. $2.8 \times 10^{-3}$
   c. $7.8 \times 10^{-2}$
   d. $1.2 \times 10^{-3}$

40. 
   a. $3.4 \times 10^{-4}$
   b. $1.0335 \times 10^4$
   c. $2 \times 10^1$
   d. $3.365 \times 10^5$

42. 170. mL;

   $\phantom{+}$ 18 mL
   $+$ 128.7 mL
   $+$ 23.45 mL
   $=$ 170.15 mL

   18 mL limits the precision to the ones place, thus the answer is rounded to 170. mL

44. Perimeter = 2×*length* + 2×*width* = 2×34.29cm + 2×26.72cm = 68.58 cm + 53.44 cm = 122.02 cm; For addition, the limiting term is the one with the smallest number of decimal places. In this case, the smallest number of decimal places is to the hundredths place, thus the final answer is reported as 122.02 cm, which contains five significant figures.

Chapter 2:    Measurements and Calculations

46.  none (10,434 is only known to the nearest whole number)

48.  a.  2.3 (the answer can only be given to two significant figures because 3.1 is only known to two significant figures)

   b.  $9.1 \times 10^2$: (the answer can only be given to the first decimal place because 4.1 is only given to the first decimal place; both numbers have the same power of ten)

   c.  $1.323 \times 10^3$: (the numbers must be first expressed as the same power of ten; $1.091 \times 10^3 + 0.221 \times 10^3 + 0.0114 \times 10^3 = 1.323 \times 10^3$)

   d.  $6.63 \times 10^{-13}$ (the answer can only be given to three significant figures because $4.22 \times 10^6$ is only given to three significant figures)

50.  a.  one (the factor of 2 has only one significant figure)

   b.  four (the sum within the parentheses will contain four significant figures)

   c.  two (based on the factor $4.7 \times 10^{-6}$ only having two significant figures)

   d.  three (based on the factor 63.9 having only three significant figures)

52.  a.  $(2.0944 + 0.0003233 + 12.22)/7.001 = (14.31)/7.001 = 2.045$

   b.  $(1.42 \times 10^2 + 1.021 \times 10^3)/(3.1 \times 10^{-1}) =$
       $(142 + 1021)/(3.1 \times 10^{-1}) = (1163)/(3.1 \times 10^{-1}) = 3752 = 3.8 \times 10^3$

   c.  $(9.762 \times 10^{-3})/(1.43 \times 10^2 + 4.51 \times 10^1) =$
       $(9.762 \times 10^{-3})/(143 + 45.1) = (9.762 \times 10^{-3})/(188.1) = 5.19 \times 10^{-5}$

   d.  $(6.1982 \times 10^{-4})^2 = (6.1982 \times 10^{-4})(6.1982 \times 10^{-4}) = 3.8418 \times 10^{-7}$

54.  an infinite number (a definition)

56.  $\dfrac{2.54 \text{ cm}}{1 \text{ in.}}$;  $3.25 \text{ in.} \times \dfrac{2.54 \text{ cm}}{1 \text{ in.}}$

   $\dfrac{1 \text{ in.}}{2.54 \text{ cm}}$;  $46.12 \text{ cm} \times \dfrac{1 \text{ in.}}{2.54 \text{ cm}}$

58.  $\dfrac{\text{lb}}{\$1.75}$;  $\$25.00 \times \dfrac{\text{lb}}{\$1.75}$

60.  a.  $2.23 \text{ m} \times \dfrac{1.0936 \text{ yd}}{1 \text{ m}} = 2.44 \text{ yd}$

   b.  $46.2 \text{ yd} \times \dfrac{1 \text{ m}}{1.0936 \text{ yd}} = 42.2 \text{ m}$

   c.  $292 \text{ cm} \times \dfrac{1 \text{ in}}{2.54 \text{ cm}} = 115 \text{ in}$

## Chapter 2: Measurements and Calculations

d. $881.2 \text{ in} \times \dfrac{2.54 \text{ cm}}{1 \text{ in}} = 2238 \text{ cm}$

e. $1043 \text{ km} \times \dfrac{1 \text{ mi}}{1.6093 \text{ km}} = 648.1 \text{ mi}$

f. $445.5 \text{ mi} \times \dfrac{1.6093 \text{ km}}{1 \text{ mi}} = 716.9 \text{ km}$

g. $36.2 \text{ m} \times \dfrac{1 \text{ km}}{1000 \text{ m}} = 0.0362 \text{ km}$

h. $0.501 \text{ km} \times \dfrac{1000 \text{ m}}{1 \text{ km}} \times \dfrac{100 \text{ cm}}{1 \text{ m}} = 5.01 \times 10^4 \text{ cm}$

62. 
a. $254.3 \text{ g} \times \dfrac{1 \text{ kg}}{1000 \text{ g}} = 0.2543 \text{ kg}$

b. $2.75 \text{ kg} \times \dfrac{1000 \text{ g}}{1 \text{ kg}} = 2750 \text{ g}$

c. $2.75 \text{ kg} \times \dfrac{2.2046 \text{ lb}}{1 \text{ kg}} = 6.06 \text{ lb}$

d. $2.75 \text{ kg} \times \dfrac{1000 \text{ g}}{1 \text{ kg}} \times \dfrac{16 \text{ oz}}{453.59 \text{ g}} = 97.0 \text{ oz}$

e. $534.1 \text{ g} \times \dfrac{1 \text{ kg}}{1000 \text{ g}} \times \dfrac{2.2046 \text{ lb}}{1 \text{ kg}} = 1.177 \text{ lb}$

f. $1.75 \text{ lb} \times \dfrac{1 \text{ kg}}{2.2046 \text{ lb}} \times \dfrac{1000 \text{ g}}{1 \text{ kg}} = 794 \text{ g}$

g. $8.7 \text{ oz} \times \dfrac{453.59 \text{ g}}{16 \text{ oz}} = 250 \text{ g}$

h. $45.9 \text{ g} \times \dfrac{16 \text{ oz}}{453.59 \text{ g}} = 1.62 \text{ oz}$

64. $2558 \text{ mi} \times \dfrac{1.6093 \text{ km}}{1 \text{ mi}} = 4117 \text{ km}$

66. $1 \times 10^{-10} \text{ m} \times \dfrac{100 \text{ cm}}{1 \text{ m}} = 1 \times 10^{-8} \text{ cm}$

$1 \times 10^{-8} \text{ cm} \times \dfrac{1 \text{ in}}{2.54 \text{ cm}} = 4 \times 10^{-9} \text{ in.}$

$1 \times 10^{-8} \text{ cm} \times \dfrac{1 \text{ m}}{100 \text{ cm}} \times \dfrac{10^9 \text{ nm}}{1 \text{ m}} = 0.1 \text{ nm}$

68. freezing

70. 273

72. Fahrenheit (F)

74. $T_C = (T_F - 32)/1.80 \qquad T_F = 1.80(T_C) + 32 \qquad T_K = T_C + 273 \qquad T_C = T_K - 273$

   a. $T_C = (T_F - 32)/1.80 = (-153°F - 32)/1.80 = (-185)/1.80 = -103°C$

   $T_K = -103°C + 273 = 170.\ K$

   b. $T_K = -153°C + 273 = 120.\ K$

   c. $T_F = 1.80(T_C) + 32 = 1.80(555°C) + 32 = 1031°F$

   d. $T_C = (T_F - 32)/1.80 = (-24°F - 32)/1.80 = -31°C$

76. $T_F = 1.80(T_C) + 32$

   a. $1.80(78.1) + 32 = 173°F$

   b. $1.80(40.) + 32 = 104°F$

   c. $1.80(-273) + 32 = -459°F$

   d. $1.80(32) + 32 = 90.°F$

78. $T_F = 1.80(T_C) + 32 \qquad T_C = (T_F - 32)/1.80 \qquad T_K = T_C + 273$

   a. $275 - 273 = 2°C$

   b. $(82 - 32)/1.80 = 28°C$

   c. $1.80(-21) + 32 = -5.8°F\ (-6°F)$

   d. $(-40 - 32)/1.80 = -40\ °C$ (Celsius and Fahrenheit temperatures are the same at –40).

80. $g/cm^3$ (g/mL)

82. 100 in.$^3$

84. Density is a *characteristic* property, which is always the same for a pure substance.

86. Ethanol is the least dense (0.785 g/cm$^3$).

88. $\text{density} = \dfrac{\text{mass}}{\text{volume}}$

   a. $m = 4.53\ kg \times \dfrac{1000\ g}{1\ kg} = 4530\ g$

   $d = \dfrac{4530\ g}{225\ cm^3} = 20.1\ g/cm^3$

Chapter 2:     Measurements and Calculations

b.   $v = 25.0 \text{ mL} \times \dfrac{1 \text{ cm}^3}{1 \text{ mL}} = 25.0 \text{ cm}^3$

$d = \dfrac{26.3 \text{ g}}{25.0 \text{ cm}^3} = 1.05 \text{ g/cm}^3$

c.   $m = 1.00 \text{ lb} \times \dfrac{1 \text{ kg}}{2.2046 \text{ lb}} \times \dfrac{1000 \text{ g}}{1 \text{ kg}} = 454 \text{ g}$

$d = \dfrac{454 \text{ g}}{500. \text{ cm}^3} = 0.907 \text{ g/cm}^3$

d.   $m = 352 \text{ mg} \times \dfrac{1 \text{ g}}{1000 \text{ mg}} = 0.352 \text{ g}$

$d = \dfrac{0.352 \text{ g}}{0.271 \text{ cm}^3} = 1.30 \text{ g/cm}^3$

90.   $4.50 \text{ L} \times \dfrac{1000 \text{ mL}}{1 \text{ L}} \times \dfrac{0.920 \text{ g}}{\text{mL}} = 4140 \text{ g}$

$375 \text{ g} \times \dfrac{\text{mL}}{0.920 \text{ g}} \times \dfrac{1 \text{ L}}{1000 \text{ L}} = 0.408 \text{ L}$

92.   $m = 3.5 \text{ lb} \times \dfrac{453.59 \text{ g}}{1 \text{ lb}} = 1.59 \times 10^3 \text{ g}$

$v = 1.2 \times 10^4 \text{ in.}^3 \times \left(\dfrac{2.54 \text{ cm}}{1 \text{ in}}\right)^3 = 1.97 \times 10^5 \text{ cm}^3$

$d = \dfrac{1.59 \times 10^3 \text{ g}}{1.97 \times 10^5 \text{ cm}^3} = 8.1 \times 10^{-3} \text{ g/cm}^3$

The material will float.

94.   e. 38 cm; The density of the box must be less than water's density of 1.0 g/cm³ in order to keep it afloat.

$2.0 \text{ lbs} \times \dfrac{453.59 \text{ g}}{1 \text{ lb}} = 907.18 \text{ g}$ (mass of box in grams)

$\dfrac{1.0 \text{ g}}{\text{cm}^3} = \dfrac{907.18 \text{ g}}{x \text{ cm}^3}$

$x = 907.18 \text{ cm}^3$ to exactly match the density of water

Thus, the volume of the box must be greater than 907.18 cm³ to make the density less than 1.0 g/cm³. Let's say:

Chapter 2: Measurements and Calculations

$Volume = length \times width \times height$

$908 \text{ cm}^3 = length \times width \times height$

$908 \text{ cm}^3 = l \times 5.0 \text{ cm} \times 5.0 \text{ cm}$

$h = 36.32 \text{ cm}$ or greater

The minimum length is thus 38 cm (closest to 36.32 cm). To test:

$908 \text{ cm}^3 = l \times w \times h$

$38 \text{ cm} \times 5.0 \text{ cm} \times 5.0 \text{ cm} = 950 \text{ cm}^3$

$\text{Density} = \dfrac{907.18 \text{ g}}{950 \text{ cm}^3} = 0.95 \text{ g/cm}^3$

96.  a.  $50.0 \text{ cm}^3 \times \dfrac{19.32 \text{ g}}{1 \text{ cm}^3} = 966 \text{ g}$

   b.  $50.0 \text{ cm}^3 \times \dfrac{7.87 \text{ g}}{1 \text{ cm}^3} = 394 \text{ g}$

   c.  $50.0 \text{ cm}^3 \times \dfrac{11.34 \text{ g}}{1 \text{ cm}^3} = 567 \text{ g}$

   d.  $50.0 \text{ cm}^3 \times \dfrac{2.70 \text{ g}}{1 \text{ cm}^3} = 135 \text{ g}$

98.  a.  $3.011 \times 10^{23} = 301,100,000,000,000,000,000,000$

   b.  $5.091 \times 10^9 = 5,091,000,000$

   c.  $7.2 \times 10^2 = 720$

   d.  $1.234 \times 10^5 = 123,400$

   e.  $4.32002 \times 10^{-4} = 0.000432002$

   f.  $3.001 \times 10^{-2} = 0.03001$

   g.  $2.9901 \times 10^{-7} = 0.00000029901$

   h.  $4.2 \times 10^{-1} = 0.42$

100.

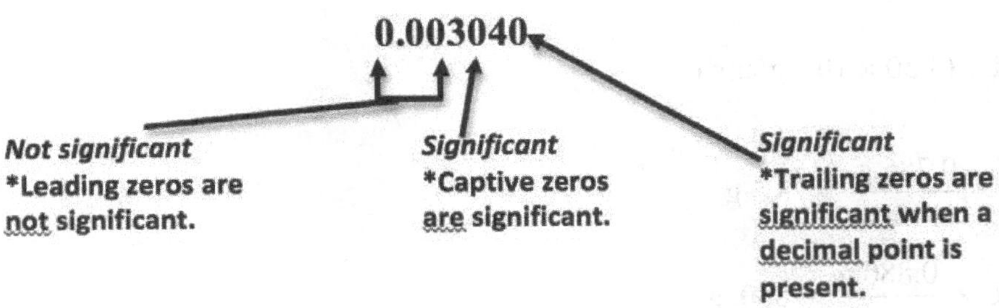

**0.003040**

*Not significant*
*Leading zeros are not significant.*

*Significant*
*Captive zeros are significant.*

*Significant*
*Trailing zeros are significant when a decimal point is present.*

Chapter 2: Measurements and Calculations

102. a. $36.2 \text{ blim} \times \dfrac{1400 \text{ kryll}}{1 \text{ blim}} = 5.07 \times 10^4 \text{ kryll}$

b. $170 \text{ kryll} \times \dfrac{1 \text{ blim}}{1400 \text{ kryll}} = 0.12 \text{ blim}$

c. $72.5 \text{ kryll}^2 \times \left(\dfrac{1 \text{ blim}}{1400 \text{ kryll}}\right)^2 = 3.70 \times 10^{-5} \text{ blim}^2$

104. Statements $a$, $c$, and $d$ are true.

Statement $a$ is true: $1.00 \text{ L} \times \dfrac{1.0567 \text{ qt}}{1 \text{ L}} = 1.06 \text{ qt}$

1.00 L is equivalent to 1.06 qt.

Statement $b$ is false: $5 \text{ ft} \times \dfrac{12 \text{ in.}}{1 \text{ ft}} = 60. \text{ in.} + 3 \text{ in.} = 63 \text{ in.}$

$63 \text{ in.} \times \dfrac{2.54 \text{ cm}}{1 \text{ in.}} \times \dfrac{1 \text{ m}}{100 \text{ cm}} = 1.6 \text{ m}$

1.6 m is taller than 1.52 m.

Statement $c$ is true: $335 \text{ g} \times \dfrac{1 \text{ lb}}{453.59 \text{ g}} = 0.739 \text{ lb}$

0.739 lb is heavier than ½ lb.

Statement $d$ is true: $\dfrac{45 \text{ mi}}{\text{hr}} \times \dfrac{1.6093 \text{ km}}{1 \text{ mi}} = 72 \text{ km/hr}$

72 km/hr is faster than 65 km/hr.

106. $1 \text{ lb} \times \dfrac{1 \text{ kg}}{2.205 \text{ lb}} \times \dfrac{2.76 \text{ euros}}{1 \text{ kg}} \times \dfrac{\$1.00}{1.44 \text{ euros}} = \$0.87$

108. °X = 1.26°C + 14

110. $d = \dfrac{36.8 \text{ g}}{10.5 \text{ L}} = 3.50 \text{ g/L}$  ($3.50 \times 10^{-3} \text{ g/cm}^3$)

112. For ethanol, $100. \text{ mL} \times \dfrac{0.785 \text{ g}}{1 \text{ mL}} = 78.5 \text{ g}$

For benzene, $1000 \text{ mL} \times \dfrac{0.880 \text{ g}}{1 \text{ mL}} = 880. \text{ g}$

total mass, 78.5 + 880. = 959 g

Chapter 2: Measurements and Calculations

114.
a. negative
b. negative
c. positive
d. zero
e. negative

116.
a. 2; positive
b. 11; negative
c. 3; positive
d. 5; negative
e. 5; positive
f. 0; zero
g. 1; negative
h. 7; negative

118.
a. 1; positive
b. 3; negative
c. 0; zero
d. 3; positive
e. 9; negative

120.
a. The decimal point must be moved five places to the left; $2.98 \times 10^{-5} = 0.0000298$.
b. The decimal point must be moved nine places to the right; $4.358 \times 10^9 = 4,358,000,000$.
c. The decimal point must be moved six places to the left; $1.9928 \times 10^{-6} = 0.0000019928$.
d. The decimal point must be moved 23 places to the right; $6.02 \times 10^{23} = 602,000,000,000,000,000,000,000$.
e. The decimal point must be moved one place to the left; $1.01 \times 10^{-1} = 0.101$.
f. The decimal point must be moved three places to the left; $7.87 \times 10^{-3} = 0.00787$.
g. The decimal point must be moved seven places to the right; $9.87 \times 10^7 = 98,700,000$.
h. The decimal point must be moved two places to the right; $3.7899 \times 10^2 = 378.99$.
i. The decimal point must be moved one place to the left; $1.093 \times 10^{-1} = 0.1093$.
j. The decimal point must be moved zero places; $2.9004 \times 10^0 = 2.9004$.
k. The decimal point must be moved four places to the left; $3.9 \times 10^{-4} = 0.00039$.
l. The decimal point must be moved eight places to the left; $1.904 \times 10^{-8} = 0.00000001904$.

122.
a. $1/10^2 = 1 \times 10^{-2}$
b. $1/10^{-2} = 1 \times 10^2$

Chapter 2:    Measurements and Calculations

c. $55/10^3 = \dfrac{5.5 \times 10^1}{1 \times 10^3} = 5.5 \times 10^{-2}$

d. $(3.1 \times 10^6)/10^{-3} = \dfrac{3.1 \times 10^6}{1 \times 10^{-3}} = 3.1 \times 10^9$

e. $(10^6)^{1/2} = 1 \times 10^3$

f. $(10^6)(10^4)/(10^2) = \dfrac{(1 \times 10^6)(1 \times 10^4)}{(1 \times 10^2)} = 1 \times 10^8$

g. $1/0.0034 = \dfrac{1}{3.4 \times 10^{-3}} = 2.9 \times 10^2$

h. $3.453/10^{-4} = \dfrac{3.453}{1 \times 10^{-4}} = 3.453 \times 10^4$

124.

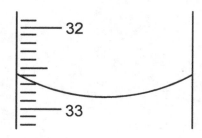

126.  $1 \text{ L} = 1 \text{ dm}^3 = 1000 \text{ cm}^3 = 1000 \text{ mL}$

128.  0.105 m

130.  They weigh the same.

$1 \text{ mg} \times \dfrac{1 \text{ g}}{1000 \text{ mg}} = 0.001 \text{ g}$

132.  $5 \times 10^{11}$ nm

$500 \text{ m} \times \dfrac{10^9 \text{ nm}}{1 \text{ m}} = 5 \times 10^{11} \text{ nm}$

134.  $v = l \times h \times w$

$0.310 \text{ m}^3 = (0.7120 \text{ m})(0.52458 \text{ m}) \times w$

$w = 0.830$ m (The answer is to three significant figures because the final volume of the box is reported to three significant figures. The other two measurements contain more significant figures and do not limit the precision of the volume.)

Chapter 2:  Measurements and Calculations

136. a. 0.000426
 b. $4.02 \times 10^{-5}$
 c. $5.99 \times 10^{6}$
 d. 400.
 e. 0.00600

138. a. 2149.6 (the answer can only be given to the first decimal place, because 149.2 is only known to the first decimal place)

 b. $5.37 \times 10^{3}$ (the answer can only be given to two decimal places because 4.34 is only known to two decimal places; moreover, since the power of ten is the same for each number, the calculation can be performed directly)

 c. Before performing the calculation, the numbers have to be converted so that they contain the same power of ten.

 $4.03 \times 10^{-2} - 2.044 \times 10^{-3} = 4.03 \times 10^{-2} - 0.2044 \times 10^{-2} = 3.83 \times 10^{-2}$ (the answer can only be given the second decimal place because $4.03 \times 10^{-2}$ is only known to the second decimal place)

 d. Before performing the calculation, the numbers have to be converted so that they contain the same power of ten.

 $2.094 \times 10^{5} - 1.073 \times 10^{6} = 2.094 \times 10^{5} - 10.73 \times 10^{5} = -8.64 \times 10^{5}$

140. a. $(2.9932 \times 10^{4})(2.4443 \times 10^{2} + 1.0032 \times 10^{1}) =$

 $(2.9932 \times 10^{4})(24.443 \times 10^{1} + 1.0032 \times 10^{1}) =$

 $(2.9932 \times 10^{4})(25.446 \times 10^{1}) = 7.6166 \times 10^{6}$

 b. $(2.34 \times 10^{2} + 2.443 \times 10^{-1})/(0.0323) =$

 $(2.34 \times 10^{2} + 0.002443 \times 10^{2})/(0.0323) =$

 $(2.34 \times 10^{2})/(0.0323) = 7.24 \times 10^{3}$

 c. $(4.38 \times 10^{-3})^{2} = 1.92 \times 10^{-5}$

 d. $(5.9938 \times 10^{-6})^{1/2} = 2.4482 \times 10^{-3}$

142.

$1 \text{ L} = 1 \text{ dm}^{3}$ (so convert all sides to dm)

$1.0 \text{ m} \times \dfrac{10 \text{ dm}}{1 \text{ m}} = 10. \text{ dm}$

$2.4 \text{ mm} \times \dfrac{1 \text{ dm}}{100 \text{ mm}} = 0.024 \text{ dm}$

$V = l \times w \times h = 10. \text{ dm} \times 0.024 \text{ dm} \times 3.9 \text{ dm} = 0.94 \text{ dm}^{3} = 0.94 \text{ L}$

Chapter 2: Measurements and Calculations

144. a. $908 \text{ oz} \times \dfrac{1 \text{ lb}}{16 \text{ oz}} \times \dfrac{1 \text{ kg}}{2.2046 \text{ lb}} = 25.7 \text{ kg}$

b. $12.8 \text{ L} \times \dfrac{1 \text{ qt}}{0.94633 \text{ L}} \times \dfrac{1 \text{ gal}}{4 \text{ qt}} = 3.38 \text{ gal}$

c. $125 \text{ mL} \times \dfrac{1 \text{ L}}{1000 \text{ mL}} \times \dfrac{1 \text{ qt}}{0.94633 \text{ L}} = 0.132 \text{ qt}$

d. $2.89 \text{ gal} \times \dfrac{4 \text{ qt}}{1 \text{ gal}} \times \dfrac{1 \text{ L}}{1.0567 \text{ qt}} \times \dfrac{1000 \text{ mL}}{1 \text{ L}} = 1.09 \times 10^4 \text{ mL}$

e. $4.48 \text{ lb} \times \dfrac{453.59 \text{ g}}{1 \text{ lb}} = 2.03 \times 10^3 \text{ g}$

f. $550 \text{ mL} \times \dfrac{1 \text{ L}}{1000 \text{ mL}} \times \dfrac{1.0567 \text{ qt}}{1 \text{ L}} = 0.58 \text{ qt}$

146. $5.3 \times 10^3 \text{ lbs} \times \dfrac{1 \text{ kg}}{2.2046 \text{ lbs}} \times \dfrac{1 \text{ metric ton}}{1000 \text{ kg}} = 2.4 \text{ metric tons}$

148. a. Celsius temperature = (175 – 32)/1.80 = 79.4°C

Kelvin temperature = 79.4 + 273 = 352 K

b. 255 – 273 = –18 °C

c. (–45 – 32)/1.80 = –43°C

d. 1.80(125) + 32 = 257°F

150. $85.5 \text{ mL} \times \dfrac{0.915 \text{ g}}{1 \text{ mL}} = 78.2 \text{ g}$

152. $m = 155 \text{ lb} \times \dfrac{453.59 \text{ g}}{1 \text{ lb}} = 7.031 \times 10^4 \text{ g}$

$v = 4.2 \text{ ft}^3 \times \left(\dfrac{12 \text{ in}}{1 \text{ ft}}\right)^3 \times \left(\dfrac{2.54 \text{ cm}}{1 \text{ in}}\right)^3 = 1.189 \times 10^5 \text{ cm}^3$

$d = \dfrac{7.031 \times 10^4 \text{ g}}{1.189 \times 10^5 \text{ cm}^3} = 0.59 \text{ g/cm}^3$

154. $T_F = 1.80(T_C) + 32$

a. 23 °F

b. 32 °F

c. –321 °F

d. –459 °F

e. 187 °F

f. −459 °F

156. a. The Mars Climate Orbiter dipped 100 km lower in the Mars atmosphere than was planned. Using the conversion factor between miles and kilometers found inside the cover of this text

$$100 \text{ km} \times \frac{1 \text{ mi}}{1.6093 \text{ km}} = 62 \text{ mi}$$

b. The aircraft required 22,300 kg of fuel, but only 22,300 lb of fuel was loaded. Using the conversion factor between pounds and kilograms found inside the cover of this text, the amount of fuel required in pounds was

$$22{,}300 \text{ kg} \times \frac{2.2046 \text{ lb}}{1 \text{ kg}} = 49{,}163 \text{ lb}$$

Therefore, $(49{,}163 - 22{,}300) = 26{,}863 = 2.69 \times 10^4$ lb additional fuel was needed.

158. $\dfrac{10^{-8} \text{ g}}{\text{L}} \times \dfrac{1 \text{ lb}}{453.59 \text{ g}} \times \dfrac{1 \text{ L}}{1.0567 \text{ qt}} \times \dfrac{4 \text{ qt}}{1 \text{ gal}} = 8 \times 10^{-11}$ lb/gal

160. 

| Number of Significant Figures | Result |
|---|---|
| 2 | 0.51 |
| 3 | 29.1 |
| 3 | 8.61 |
| 3 | 1.89 |
| 4 | 134.6 |
| 3 | 14.4 |

162. $1.25 \text{ mi} \times \dfrac{1.6093 \text{ km}}{1 \text{ mi}} \times \dfrac{1000 \text{ m}}{1 \text{ km}} = 2011.625 \text{ m}$

60 sec + 59.2 sec = 119.2 sec

$\dfrac{2011.625 \text{ m}}{119.2 \text{ s}} = 16.9$ m/s

164. $T_C = (T_F - 32)/1.80$

$T_C = (134 - 32)/1.80 = 56.7°C$

Since the temperature is higher than the melting point (44°C), phosphorus would be a liquid.

Chapter 2:   Measurements and Calculations

166.    $69 \text{ pm} \times \dfrac{1 \text{ m}}{10^{12} \text{ pm}} \times \dfrac{100 \text{ cm}}{1 \text{ m}} = 6.9 \times 10^{-9} \text{ cm}$

$V = \tfrac{4}{3}\pi(6.9 \times 10^{-9} \text{ cm})^3 = 1.4 \times 10^{-24} \text{ cm}^3$

$d = \dfrac{mass}{volume} = \dfrac{3.35 \times 10^{-23} \text{ g}}{1.4 \times 10^{-24} \text{ cm}^3} = 24 \text{ g/cm}^3$

# CHAPTER 3

# Matter

2. intermolecular forces

4. liquids

6. gaseous

8. The *stronger* the inter-particle forces, the more rigid is the sample overall.

10. Gases are easily compressed into smaller volumes, whereas solids and liquids are not. Because a gaseous sample consists mostly of empty space, the gas particles are pushed closer together when pressure is applied to a gas.

12. chemical change; New products are formed. The reactants are water molecules, which undergoes a chemical reaction to produce hydrogen and oxygen molecules. These are chemically different than water molecules.

14. Magnesium is malleable and ductile.

16. (d); The identity of the molecules that make up (a), (b) and (c) does not change, just the rearrangement.

18. 
   a. physical; the iron is only being heated.
   b. chemical; the sugars in the marshmallow are being reduced to carbon.
   c. chemical; most strips contain a peroxide which decomposes.
   d. chemical; the bleach oxidizes dyes in the fabric.
   e. physical; evaporation is only a change of state.
   f. physical; the salt is only modifying the physical properties of the solution, not undergoing a chemical reaction.
   g. chemical; the drain cleaner breaks bonds in the hair.
   h. physical; students will most likely reply that this is a physical change since the perfume is evaporating; the sensation of smell, however, depends on chemical processes.
   i. physical; the sublimation is only a change of state.
   j. physical; the wood is only being physically divided into smaller pieces.
   k. chemical; the cellulose in the wood is reacting with oxygen gas

20. Compounds consist of two or more elements combined together chemically in a fixed composition, no matter what their source may be. For example, water on earth consists of molecules containing one oxygen atom and two hydrogen atoms. Water on Mars (or any other planet) has the same composition.

22. compounds

24. HCl; Elements (e.g., He, $F_2$, $S_8$) cannot be broken down into other substances by chemical means. HCl is a compound because it can be broken down into hydrogen and chlorine.

Chapter 3:   Matter

26. Given that the product of the process is no longer attracted by the magnet, this strongly suggests that the iron has been converted to an iron/sulfur compound—a pure substance.

28. c. air (in this room) and d. gasoline (for a car) are both examples of homogeneous mixtures; A chocolate chip cookie is an example of a heterogeneous mixture. Iodine crystals are an example of an element. Sucrose is an example of a compound.

30. a.   mixture
    b.   mixture
    c.   mixture
    d.   pure substance

32. Concrete is a mixture: the various components of the particular concrete are still distinguishable within the concrete if examined closely.

34. Consider a mixture of salt (sodium chloride) and sand. Salt is soluble in water, sand is not. The mixture is added to water and stirred to dissolve the salt and is then filtered. The salt solution passes through the filter; the sand remains on the filter. The water can then be evaporated from the salt.

36. The chemical identities of the components of the mixture are not changed by filtration or distillation: the various components are separated by physical, not chemical, means.

38. a.   compound; pure substance
    b.   element; pure substance
    c.   homogeneous mixture

40. Because vaporized water is still the *same substance* as solid water, no chemical reaction has occurred. Sublimation is a physical change.

42. False. No reaction has taken place. The substances are merely separating, not changing into different substances. This is an example of a heterogeneous mixture.

44. (b); $P_4$ is an element. Dissolving sugar in water is not a chemical change (both the sugar and water are still intact). NaCl is a compound.

46. physical

48. pure substance; compound; element

50. The correct answer is *d. sodium* (element), *sodium chloride* (compound), *salt water* (mixture).

    Incorrect:   a. copper (element), silicon dioxide (compound), copper(II) sulfate (compound)

    b. hydrogen (element), carbon dioxide (compound), water (compound)

    c. chili (mixture), pizza (mixture), steak (mixture)

    e. nitrogen (element), argon (element), air (mixture)

52. a. Air is a *homogeneous mixture* of gases because it generally contains the same ratio of gaseous substances from one region to another.

54. False. The substances in the mixture do not always combine to form a new product. Mixtures can be separated into pure substances, but this is not a chemical reaction.

56. $O_2$ and $P_4$ are both still elements, even though the ordinary forms of these elements consist of molecules containing more than one atom (but all atoms in each respective molecule are the

same). $P_2O_5$ is a compound, because it is made up of two or more different elements (not all the atoms in the $P_2O_5$ molecule are the same).

58. Assuming there is enough water present in the mixture to have dissolved all the salt, filter the mixture to separate out the sand from the mixture. Then distill the filtrate (consisting of salt and water), which will boil off the water, leaving the salt.

60. The most obvious difference is the physical states: water is a liquid under room conditions, hydrogen and oxygen are both gases. Hydrogen is flammable. Oxygen supports combustion. Water does neither.

62. 
   a. False. A spoonful of sugar is a compound (sucrose, $C_{12}H_{22}O_{11}$).
   b. False. Element and compounds are pure substances.
   c. True.
   d. False. Gasoline is a mixture.
   e. True.

64. (a), (d)

# CUMULATIVE REVIEW

# Chapters 1–3

2. By now, after having covered three chapters in this book, it is hoped that you have adopted an "active" approach to your study of chemistry. You may have discovered (perhaps through a disappointing grade on a quiz (though we hope not), that you really have to get involved with chemistry. You can't just sit and take notes, or just look over the solved examples in the textbook. You have to learn to solve problems. You have to learn how to interpret problems, and how to reduce them to the simple mathematical relationships you have studied. Whereas in some courses you might get by on just giving back on exams the facts or ideas presented in class, in chemistry you have to be able to extend and synthesize what has been discussed and to apply the material to new situations. Don't get discouraged if this is difficult at first: it's difficult for everyone at first.

4. It is difficult sometimes for students (especially beginning students) to understand why certain subjects are required for a given college major. The faculty of your major department, however, have collectively many years of experience in the subject in which you have chosen to specialize. They really do know what courses will be helpful to you in the future. They may have had trouble with the same courses that now give you trouble, but they realize that all the work will be worth it in the end. Some courses you take, particularly in your major field itself, have obvious and immediate utility. Other courses, often times chemistry included, are provided to give you a general background knowledge, which may prove useful in understanding your own major or other subjects related to your major. In perhaps a burst of bravado, chemistry has been called "the central science" by one team of textbook authors. This moniker is very true however: in order to understand biology, physics, nutrition, farming, home economics, or whatever (it helps to have a general background in chemistry).

6. Whenever a scientific measurement is made, we always employ the instrument or measuring device we are using to the limits of its precision. On a practical basis, this usually means that we *estimate* our reading of the last significant figure of the measurement. An example of the uncertainty in the last significant figure is given for measuring the length of a pin in the text in Figure 2.5. Scientists appreciate the limits of experimental techniques and instruments, and always assume that the last digit in a number representing a measurement has been estimated. Because the last significant figure in every measurement is assumed to be estimated, it is never possible to exclude uncertainty from measurements. The best we can do is to try to improve our techniques and instruments so that we get more significant figures for our measurements.

8. Dimensional analysis is a method of problem solving that pays particular attention to the units of measurements and uses these units as if they were algebraic symbols that multiply, divide, and cancel. Consider the following example. A dozen eggs costs $1.25. Suppose we want to know how much one egg costs, and also how much three dozens of eggs will cost. To solve these problems, we need to make use of two equivalence statements:

   1 dozen eggs = 12 eggs

   1 dozen eggs = $1.25

   The first of these equivalence statements is obvious: everyone knows that 12 eggs is "equivalent" to one dozen. The second statement also expresses an equivalence: if you give the grocer $1.25, he or she will give you a dozen eggs. From these equivalence statements, we can construct the conversion factors we need to answer the two questions. For the first question (what does one egg cost) we can set up the calculation as follows

$$\frac{\$1.25}{12 \text{ eggs}} = \$0.104 = \$0.10$$

as the cost of one egg. Similarly, for the second question (the cost of 3 dozens eggs), we can set up the conversion as follows

$$3 \text{ dozens} \times \frac{\$1.25}{1 \text{ dozen}} = \$3.75$$

as the cost of three dozens eggs. See Section 2.6 of the text for how we construct conversion factors from equivalence statements.

10. Defining what scientists mean by "matter" often seems circular to students. Scientists say that matter is something that "has mass and occupies space", without ever really explaining what it means to "have mass" or to "occupy space"! The concept of matter is so basic and fundamental, that it becomes difficult to give a good textbook definition other than to say that matter is the "stuff" of which everything is made. Matter can be classified and subdivided in many ways, depending on what we are trying to demonstrate.

On the most fundamental basis, all matter is composed of tiny particles (such as protons, electrons, neutrons, and the other subatomic particles). On one higher level, these tiny particles are combined in a systematic manner into units called atoms. Atoms, in turn, may be combined to constitute molecules. And finally, large groups of molecules may be placed together to form a bulk sample of substance that we can see.

Matter can also be classified as to the physical state a particular substance happens to take. Some substances are solids, some are liquids, and some are gases. Matter can also be classified as to whether it is a pure substance (one type of molecule) or a mixture (more than one type of molecule), and furthermore whether a mixture is homogeneous or heterogeneous.

12. Chemists tend to give a functional definition of what they mean by an "element": an element is a fundamental substance that cannot be broken down into any simpler substances by chemical methods. Compounds, on the other hand, can be broken down into simpler substances (the elements of which the compound is composed). For example, sulfur and oxygen are both elements (sulfur occurs as $S_8$ molecules and oxygen as $O_2$ molecules). When sulfur and oxygen are placed together and heated, the compound sulfur dioxide ($SO_2$) forms. When we analyze the sulfur dioxide produced, we notice that each and every molecule consists of one sulfur atom and two oxygen atoms, and on a mass basis, consists of 50% each of sulfur and oxygen. We describe this by saying that sulfur dioxide has a constant composition. The fact that a given compound has constant composition is usually expressed in terms of the mass percentages of the elements present in the compound. The reason the mass percentages are constant is because of a constant number of atoms of each type present in the compound's molecules. If a scientist anywhere in the universe analyzed sulfur dioxide, he or she would find the same composition: if a scientist finds something that does not have the same composition, then the substance cannot be sulfur dioxide.

14. a. The decimal point must be moved six places to the left: $4.861903 \times 10^6$

Review: Chapters 1, 2, and 3

    b.    The decimal point must be moved two places to the right: 381.36

    c.    The decimal point must be moved three places to the left: 0.0051

    d.    The decimal point must be moved four places to the right: $7.44 \times 10^{-4}$

    e.    The arithmetic must be performed and then the exponents combined: $4.615 \times 10^3$

    f.    The arithmetic must be performed and then the exponents combined: $1.527 \times 10^{-10}$

16.    a.    two (based on the factor of 2.1 in the denominator)

    b.    two (based on the factor of 5.2 in the numerator)

    c.    three (one before the decimal point, and two after the decimal point)

    d.    three (based on the sum of 5.338 and 2.11)

    e.    one (based on 9 only having one significant figure)

    f.    two (based on the sum of 4.2005 and 2.7)

    g.    two (based on the factor of 0.15)

    h.    three (two before the decimal point, and one after the decimal point)

18.    density = mass/volume    mass = volume × density    volume = mass/density

    a.    $\text{density} = \dfrac{78.5 \text{ g}}{100. \text{ mL}} = 0.785 \text{ g/mL}$

    b.    $\text{volume} = \text{mass/density} = \dfrac{1.590 \text{ kg} \times \dfrac{1000 \text{ g}}{1 \text{ kg}}}{0.785 \text{ g/mL}} = 2025 \text{ mL} = 2.03 \text{ L}$

    c.    $\text{mass} = \text{volume} \times \text{density} = 1.35 \text{ L} \times \dfrac{1000 \text{ mL}}{1 \text{ L}} \times \dfrac{0.785 \text{ g}}{1 \text{ mL}} = 1060 \text{ g} = 1.06 \text{ kg}$

    d.    $\text{volume} = \text{mass/density} = \dfrac{25.2 \text{ g}}{2.70 \text{ g/cm}^3} = 9.33 \text{ cm}^3$

    e.    $\text{volume} = 12.0 \text{ cm} \times 2.5 \text{ cm} \times 2.5 \text{ cm} = 75 \text{ cm}^3$

           $\text{mass} = \text{volume} \times \text{density} = 75 \text{ cm}^3 \times 2.70 \text{ g/cm}^3 = 202.5 \text{ g} = 2.0 \times 10^2 \text{ g}$

# CHAPTER 4

# Chemical Foundations: Elements, Atoms, and Ions

2. Robert Boyle

4. oxygen, carbon, hydrogen

6. a. Trace elements are those elements which are present in only tiny amounts in the body, but are critical for many bodily processes and functions.

   b. Answer depends on your choice of elements

8. Sometimes the symbol for an element is based on its common name in another language. This is true for many of the more common metals since their existence was known to the ancients: some examples are iron, sodium, potassium, silver, and tin (the symbols come from their name in Latin); tungsten (the symbol comes from its name in German).

10. a. neon
    b. nickel
    c. nitrogen
    d. nobelium
    e. neptunium
    f. niobium
    g. neodymium

12. Zr   zirconium
    Cs   cesium
    Se   selenium
    Au   gold
    Ce   cerium

14. B: barium, Ba; berkelium, Bk; beryllium, Be; bismuth, Bi; bohrium, Bh; boron, B; bromine, Br

    N: neodymium, Nd; neon, Ne; neptunium, Np; nickel, Ni; niobium, Nb; nitrogen, N; nobelium, No

    P: palladium, Pd; phosphorus, P; platinum, Pt; plutonium, Pu; polonium, Po; potassium, K; praseodymium, Pr; promethium, Pm; protactinium, Pa

    S: samarium, Sm; scandium, Sc; seaborgium, Sg; selenium, Se; silicon, Si; silver, Ag; sodium, Na; strontium, Sr; sulfur, S

16. a. Elements are made of tiny particles called atoms.

    b. All the atoms of a given element are identical

    c. The atoms of a given element are different from those of any other element.

    d. A given compound always has the same numbers and types of atoms.

Chapter 4: Elements, Atoms, and Ions

  e. Atoms are neither created nor destroyed in chemical processes. A chemical reaction simply changes the way the atoms are grouped together.

18. According to Dalton, all atoms of the same element are *identical*; in particular, every atom of a given element has the same *mass* as every other atom of that element. If a given compound always contains the *same relative numbers* of atoms of each kind, and those atoms always have the *same masses*, then it follows that the compound made from those elements would always contain the same relative masses of its elements.

20. a. $CO_2$      d. $H_2SO_4$

  b. $CO$       e. $BaCl_2$

  c. $CaCO_3$     f. $Al_2S_3$

22. False. Rutherford's bombardment experiments with metal foil suggested that the alpha particles were being deflected by coming near a *dense, positively charged* atomic nucleus.

24. protons

26. neutron; electron

28. the electrons; outside the nucleus; Because they are located in the exterior regions of the atom, it is the electrons of an atom that most interact with other atoms and are therefore most responsible for the atom's chemical behavior.

30. False. The mass number represents the total number of protons and neutrons in the nucleus.

32. Neutrons are uncharged and contribute only to the mass.

34. James Chadwick

36. (*a*) and (*b*); Isotopes are atoms with the same number of protons but different numbers of neutrons. In a nuclide symbol, the bottom number represents the number of protons in the atom. The top number represents the mass number, which is the sum of the number of neutrons and number of protons in the atom. Thus, both *a* and *b* have the same number of protons (10 protons) but different numbers of neutrons (*a* contains 10 neutrons and *b* contains 12 neutrons).

38. a. $^{54}_{26}Fe$

  b. $^{56}_{26}Fe$

  c. $^{57}_{26}Fe$

  d. $^{14}_{7}N$

  e. $^{15}_{7}N$

  f. $^{15}_{7}N$

40. The relative amounts of $^2H$ and $^{18}O$ in a person's hair, compared to other isotopes of these elements, vary significantly from region to region in the United States and is related to the isotopic abundances in the drinking water in a region.

42.

| Name | Symbol | Atomic Number | Mass Number | Number of neutrons |
|---|---|---|---|---|
| oxygen | $^{17}_{8}O$ | 8 | 17 | 9 |
| oxygen | $^{17}_{8}O$ | 8 | 17 | 9 |
| neon | $^{20}_{10}Ne$ | 10 | 20 | 10 |
| iron | $^{56}_{26}Fe$ | 26 | 56 | 30 |
| plutonium | $^{244}_{94}Pu$ | 94 | 244 | 150 |
| mercury | $^{202}_{80}Hg$ | 80 | 202 | 122 |
| cobalt | $^{59}_{27}Co$ | 27 | 59 | 32 |
| nickel | $^{56}_{28}Ni$ | 28 | 56 | 28 |
| fluorine | $^{19}_{9}F$ | 9 | 19 | 10 |
| chromium | $^{50}_{24}Cr$ | 24 | 50 | 26 |

44. Elements with similar chemical properties are aligned *vertically* in families known as *groups*.

46. Metallic elements are found toward the *left* and *bottom* of the periodic table. There are far more metallic elements than there are nonmetals.

48. The gaseous nonmetallic elements are hydrogen, nitrogen, oxygen, fluorine, chlorine, plus all the group 8 elements (noble gases). There are no gaseous metallic elements under room conditions.

50. metalloids or semimetals

52.  a.  fluorine, chlorine, bromine, iodine, astatine

  b.  lithium, sodium, potassium, rubidium, cesium, francium

  c.  beryllium, magnesium, calcium, strontium, barium, radium

  d.  helium, neon, argon, krypton, xenon, radon

54. Arsenic, atomic number 33, is located on the dividing line between the metallic elements and the non-metallic elements, and is therefore classified as a metalloid. Arsenic is in Group 5 of the periodic table, whose other principal members are N, P, Sb, and Bi.

56. Most of the elements are too reactive to be found in the uncombined form in nature and are found only in compounds.

58. These elements are found *uncombined* in nature and do not readily react with other elements. For many years it was thought that these elements formed no compounds at all, although this has now been shown to be untrue.

60. diatomic gases: $H_2$, $N_2$, $O_2$, $Cl_2$, and $F_2$

  monatomic gases: He, Ne, Kr, Xe, Rn, and Ar

62. chlorine

64. graphite

66. electrons

68. 2–

70. *-ide*

72. False. $N^{3-}$ contains 7 protons and 10 electrons. $P^{3-}$ contains 15 protons and 18 electrons.

Chapter 4: Elements, Atoms, and Ions

74. number of protons = 8; number of electrons = 10; number of neutrons = 9
76. 
   a. two electrons gained
   b. three electrons gained
   c. three electrons lost
   d. two electrons lost
   e. one electron lost
   f. two electrons lost.
78. 
   a. $P^{3-}$
   b. $Ra^{2+}$
   c. $At^{-}$
   d. no ion
   e. $Cs^{+}$
   f. $Se^{2-}$
80. Sodium chloride is an ionic compound, consisting of $Na^+$ and $Cl^-$ ions. When NaCl is dissolved in water, these ions are set free and can move independently to conduct the electric current.
82. The total number of positive charges must equal the total number of negative charges so that there will be *no net charge* on the crystals of an ionic compound. A macroscopic sample of compound must ordinarily not have any net charge.
84. 
   a. CsI, $BaI_2$, $AlI_3$
   b. $Cs_2O$, BaO, $Al_2O_3$
   c. $Cs_3P$, $Ba_3P_2$, AlP
   d. $Cs_2Se$, BaSe, $Al_2Se_3$
   e. CsH, $BaH_2$, $AlH_3$
86. 
   a. 7; halogens
   b. 8; noble gases
   c. 2; alkaline earth elements
   d. 2; alkaline earth elements
   e. 4
   f. 6; (the members of group 6 are sometimes called the chalcogens)
   g. 8; noble gases
   h. 1; alkali metals
88. (*b*); Dalton's atomic theory stated that all atoms of a given element are identical.
90. Most of the mass of an atom is concentrated in the nucleus: the *protons* and *neutrons* that constitute the nucleus have similar masses, and these particles are nearly two thousand times heavier than electrons. The chemical properties of an atom depend on the number and location of

Chapter 4: Elements, Atoms, and Ions

the *electrons* it possesses. Electrons are found in the outer regions of the atom and are the particles most likely to be involved in interactions between atoms.

92. $C_6H_{12}O_6$

94. 
a. 29 protons; 34 neutrons; 29 electrons
b. 35 protons; 45 neutrons; 35 electrons
c. 12 protons; 12 neutrons; 12 electrons

96. The chief use of gold in ancient times was as *ornamentation*, whether in statuary or in jewelry. Gold possesses an especially beautiful luster, and because it is relatively soft and malleable, it could be worked finely by artisans. Among the metals, gold is particularly inert to attack by most substances in the environment.

98. 
a. 36
b. 36
c. 21
d. 36
e. 80
f. 27

100. (*e*); B, Si, and Ge are considered metalloids or semimetals.

102. The metal ion is $Cu^{2+}$. Since the metal ion has 27 electrons and contains a 2+ charge, this means that it has two less electrons as compared to protons. Therefore the number of protons is 29. The number of protons is also the atomic number, identifying the metal ion as copper. Mass number = 29 $p^+$ + 34 n = 63

104. 
a. $CO_2$
b. $AlCl_3$
c. $HClO_4$
d. $SCl_6$

106. 
a. $^{13}_{6}C$
b. $^{13}_{6}C$
c. $^{13}_{6}C$
d. $^{44}_{19}K$
e. $^{41}_{20}Ca$
f. $^{35}_{19}K$

## Chapter 4: Elements, Atoms, and Ions

108.

| Symbol | Protons | Neutrons | Mass Number |
|---|---|---|---|
| $^{41}_{20}$Ca | 20 | 21 | 41 |
| $^{55}_{25}$Mn | 25 | 30 | 55 |
| $^{109}_{47}$Ag | 47 | 62 | 109 |
| $^{45}_{21}$Sc | 21 | 24 | 45 |

110. **Cu-63**: 29 protons, 29 electrons, 34 neutrons, $^{63}_{29}$Cu

**Cu-65**: 29 protons, 29 electrons, 36 neutrons, $^{65}_{29}$Cu

112. **tin**: Sn
**beryllium**: Be
**hydrogen**: H
**chlorine**: Cl
**radium**: Ra
**xenon**: Xe
**zinc**: Zn
**oxygen**: O

114.

| Atom | G or L | Ion |
|---|---|---|
| O | G | $O^{2-}$ |
| Mg | L | $Mg^{2+}$ |
| Rb | L | $Rb^{+}$ |
| Br | G | $Br^{-}$ |
| Cl | G | $Cl^{-}$ |

116.

| Atom/Ion | Protons | Neutrons | Electrons |
|---|---|---|---|
| $^{120}_{50}$Sn | 50 | 70 | 50 |
| $^{25}_{12}$Mg$^{2+}$ | 12 | 13 | 10 |
| $^{56}_{26}$Fe$^{2+}$ | 26 | 30 | 24 |
| $^{79}_{34}$Se | 34 | 45 | 34 |
| $^{35}_{17}$Cl | 17 | 18 | 17 |
| $^{63}_{29}$Cu | 29 | 34 | 29 |

# CHAPTER 5

# Nomenclature

2. A binary compound contains only two elements: the major types of binary compounds are *ionic* (compounds that contain a metal and a nonmetal) and *nonionic* (compounds containing two nonmetals).

4. anion (negative ion)

6. Sodium chloride consists of $Na^+$ ions and $Cl^-$ ions in an extended crystal lattice array. No discrete NaCl pairs are present.

8. Roman numeral

10. a. lithium iodide
    b. magnesium fluoride
    c. strontium oxide
    d. aluminum bromide
    e. calcium sulfide
    f. sodium oxide

12. a. incorrect; copper(II) chloride
    b. correct
    c. correct
    d. incorrect; calcium sulfide
    e. correct

14. a. As the iodide ion has a 1– charge, the iron ion must have a 3+ charge: the name is iron(III) iodide.
    b. As the chloride ion has a 1– charge, the manganese ion must have a 2+ charge: the name is manganese(II) chloride.
    c. As the oxide ion has a 2– charge, the mercury ion must have a 2+ charge: the name is mercury(II) oxide.
    d. As the sulfide ion has a 2– charge, each copper ion must have a 1+ charge: the name is copper(I) sulfide.
    e. As the oxide ion has a 2– charge, the cobalt ion must have a 2+ charge: the name is cobalt(II) oxide.
    f. As the bromide ion has a 1– charge, the tin ion must have a 4+ charge: the name is tin(IV) bromide.

Chapter 5:    Nomenclature

16.  a.  As each chloride ion has a 1– charge, the cobalt ion must have a 2+ charge: the name is cobalt*ous* chloride.

b.  As each bromide ion has a 1– charge, the chromium ion must have a 3+ charge: the name is chrom*ic* bromide.

c.  As each oxide ion has a 2– charge, the lead ion must have a 2+ charge: the name is plumb*ous* oxide.

d.  As each oxide ion has a 2– charge, the tin ion must have a 4+ charge: the name is stann*ic* oxide.

e.  As the oxide ion has a 2– charge, the cobalt ion must have a 3+ charge: the name is cobalt*ic* oxide.

f.  As the chloride ion has a 1– charge, the iron ion must have a 3+ charge: the name is ferr*ic* chloride.

18.  Remember that for this type of compound of nonmetals, numerical prefixes are used to indicate how many of each type of atom are present. However, if only one atom of the first element mentioned in the compound is present in a molecule, the prefix *mono–* is not needed.

a.  chlorine pentafluoride

b.  xenon dichloride

c.  selenium dioxide

d.  dinitrogen trioxide

e.  diiodine hexachloride

f.  carbon disulfide

20.  $Na_2O$: sodium oxide; $N_2O$: dinitrogen monoxide; For $Na_2O$, the compound contains a metal and a nonmetal in which the charges must balance. When forming this compound, Na always forms a 1+ charge and oxygen always forms a 2– charge. Therefore, the prefixes are not needed. For $N_2O$, the compound contains only nonmetals and the charges do not have to balance. Therefore prefixes are needed to tell us how many of each atom are present.

22.  a.  radium chloride – ionic

b.  selenium dichloride – nonionic

c.  phosphorus trichloride – nonionic

d.  sodium phosphide – ionic

e.  manganese(II) fluoride (or manganous fluoride) – ionic

f.  zinc oxide – ionic

24.  oxyanion

26.  For a series of oxyanions, the prefix *hypo–* is used for the anion with the fewest oxygen atoms, and the prefix *per–* is used for the anion with the most oxygen atoms.

28.  $IO^-$    hypoiodite

$IO_2^-$    iodite

IO$_3^-$ iodate
IO$_4^-$ periodate

30. a. CN$^-$
    b. CO$_3^{2-}$
    c. HCO$_3^-$
    d. C$_2$H$_3$O$_2^-$

32. CN$^-$ cyanide
    CO$_3^{2-}$ carbonate
    HCO$_3^-$ hydrogen carbonate (or bicarbonate)
    C$_2$H$_3$O$_2^-$ acetate

34. a. ammonium
    b. dihydrogen phosphate
    c. sulfate
    d. hydrogen sulfite (also called *bi*sulfite)
    e. perchlorate
    f. iodate

36. a. sodium permanganate
    b. aluminum phosphate
    c. chromium(II) carbonate, chromous carbonate
    d. calcium hypochlorite
    e. barium carbonate
    f. calcium chromate

38. oxygen (commonly referred to as *oxy*acids)

40. a. hypochlorous acid
    b. sulfurous acid
    c. bromic acid
    d. hypoiodous acid
    e. perbromic acid
    f. hydrosulfuric acid
    g. hydroselenic acid
    h. phosphorous acid

Chapter 5:     Nomenclature

42. 
   a. MgF$_2$
   b. FeI$_3$
   c. HgS
   d. Ba$_3$N$_2$
   e. PbCl$_2$
   f. SnF$_4$
   g. Ag$_2$O
   h. K$_2$Se

44. 
   a. P$_2$O
   b. SO$_2$
   c. P$_2$O$_5$
   d. CCl$_4$
   e. NBr$_3$
   f. SiF$_4$
   g. SCl$_2$

46. 
   a. NH$_4$C$_2$H$_3$O$_2$
   b. Fe(OH)$_2$
   c. Co$_2$(CO$_3$)$_3$
   d. BaCr$_2$O$_7$
   e. PbSO$_4$
   f. KH$_2$PO$_4$
   g. Li$_2$O$_2$
   h. Zn(ClO$_3$)$_2$

48. 
   a. HCN
   b. HNO$_3$
   c. H$_2$SO$_4$
   d. H$_3$PO$_4$
   e. HClO or HOCl
   f. HBr
   g. HBrO$_2$
   h. HF

50. 
   a. Ca(HSO$_4$)$_2$
   b. Zn$_3$(PO$_4$)$_2$

c. Fe(ClO$_4$)$_3$

d. Co(OH)$_3$

e. K$_2$CrO$_4$

f. Al(H$_2$PO$_4$)$_3$

g. LiHCO$_3$

h. Mn(C$_2$H$_3$O$_2$)$_2$

i. MgHPO$_4$

j. CsClO$_2$

k. BaO$_2$

l. NiCO$_3$

52. A moist paste of NaCl would contain Na$^+$ and Cl$^-$ ions in solution, and would serve as a *conductor* of electrical impulses.

54. H → H$^+$ (hydrogen ion: a cation) + e$^-$

H + e$^-$ → H$^-$ (hyd*ride* ion: an anion)

56. missing oxyanions:  IO$_3^-$; ClO$_2^-$

missing oxyacids:  HClO$_4$; HClO; HBrO$_2$

58. a. gold(III) bromide, auric bromide

b. cobalt(III) cyanide, cobaltic cyanide

c. magnesium hydrogen phosphate

d. diboron hexahydride (diborane is its common name)

e. ammonia

f. silver(I) sulfate (usually called silver sulfate)

g. beryllium hydroxide

60. (b); iron(II) oxide has the formula FeO

62. a. M(C$_2$H$_3$O$_2$)$_2$

b. M(MnO$_4$)$_2$

c. MO

d. MHPO$_4$

e. M(OH)$_2$

f. M(NO$_2$)$_2$

64. a. The metal ion is Mn$^{2+}$. Since the metal ion has 23 electrons and contains a 2+ charge, this means that it has two less electrons as compared to protons. Therefore the number of protons is 25. The number of protons is also the atomic number, identifying the metal ion as manganese.

## Chapter 5: Nomenclature

b. The halogen ion is Cl⁻ with 18 electrons. The number of protons is 17, identifying the element as chlorine. Halogens form a 1– charge when bonding with a metal to form an ionic compound, thus the ion has one more electron as compared to protons.

c. Since the chloride ion has a 1– charge and the manganese ion has a 2+ charge, the formula is $MnCl_2$ and is named manganese(II) chloride (or manganous chloride). The charge on manganese must be specified using the Roman numeral.

66.  

| | | | | | |
|---|---|---|---|---|---|
| $Ca(NO_3)_2$ | $CaSO_4$ | $Ca(HSO_4)_2$ | $Ca(H_2PO_4)_2$ | $CaO$ | $CaCl_2$ |
| $Sr(NO_3)_2$ | $SrSO_4$ | $Sr(HSO_4)_2$ | $Sr(H_2PO_4)_2$ | $SrO$ | $SrCl_2$ |
| $NH_4NO_3$ | $(NH_4)_2SO_4$ | $NH_4HSO_4$ | $NH_4H_2PO_4$ | $(NH_4)_2O$ | $NH_4Cl$ |
| $Al(NO_3)_3$ | $Al_2(SO_4)_3$ | $Al(HSO_4)_3$ | $Al(H_2PO_4)_3$ | $Al_2O_3$ | $AlCl_3$ |
| $Fe(NO_3)_3$ | $Fe_2(SO_4)_3$ | $Fe(HSO_4)_3$ | $Fe(H_2PO_4)_3$ | $Fe_2O_3$ | $FeCl_3$ |
| $Ni(NO_3)_2$ | $NiSO_4$ | $Ni(HSO_4)_2$ | $Ni(H_2PO_4)_2$ | $NiO$ | $NiCl_2$ |
| $AgNO_3$ | $Ag_2SO_4$ | $AgHSO_4$ | $AgH_2PO_4$ | $Ag_2O$ | $AgCl$ |
| $Au(NO_3)_3$ | $Au_2(SO_4)_3$ | $Au(HSO_4)_3$ | $Au(H_2PO_4)_3$ | $Au_2O_3$ | $AuCl_3$ |
| $KNO_3$ | $K_2SO_4$ | $KHSO_4$ | $KH_2PO_4$ | $K_2O$ | $KCl$ |
| $Hg(NO_3)_2$ | $HgSO_4$ | $Hg(HSO_4)_2$ | $Hg(H_2PO_4)_2$ | $HgO$ | $HgCl_2$ |
| $Ba(NO_3)_2$ | $BaSO_4$ | $Ba(HSO_4)_2$ | $Ba(H_2PO_4)_2$ | $BaO$ | $BaCl_2$ |

68. $(NH_4)_3PO_4$

70. iodine (solid), bromine (liquid), fluorine and chlorine (gases)

72. 1+

74. 2–

76.  
a. $Al(13e^-) \rightarrow Al^{3+}(10e^-) + 3e^-$

b. $S(16e^-) + 2e^- \rightarrow S^{2-}(18e^-)$

c. $Cu(29e^-) \rightarrow Cu^+(28e^-) + e^-$

d. $F(9e^-) + e^- \rightarrow F^-(10e^-)$

e. $Zn(30e^-) \rightarrow Zn^{2+}(28e^-) + 2e^-$

f. $P(15e^-) + 3e^- \rightarrow P^{3-}(18e^-)$

78.  
a. Two 1+ ions are needed to balance a 2– ion, so the formula must have two $Na^+$ ions for each $S^{2-}$ ion: $Na_2S$.

b. One 1+ ion exactly balances a 1– ion, so the formula should have an equal number of $K^+$ and $Cl^-$ ions: $KCl$.

c. One 2+ ion exactly balances a 2– ion, so the formula must have an equal number of $Ba^{2+}$ and $O^{2-}$ ions: $BaO$.

d. One 2+ ion exactly balances a 2– ion, so the formula must have an equal number of $Mg^{2+}$ and $Se^{2-}$ ions: $MgSe$.

e. One 2+ ion requires two 1– ions to balance charge, so the formula must have twice as many $Br^-$ ions as $Cu^{2+}$ ions: $CuBr_2$.

f. One 3+ ion requires three 1– ions to balance charge, so the formula must have three times as many $I^-$ ions as $Al^{3+}$ ions: $AlI_3$.

Chapter 5:   Nomenclature

g.  Two 3+ ions give a total of 6+, whereas three 2– ions will give a total of 6–. The formula then should contain two $Al^{3+}$ ions and three $O^{2-}$ ions: $Al_2O_3$.

h.  Three 2+ ions are required to balance two 3– ions, so the formula must contain three $Ca^{2+}$ ions for every two $N^{3-}$ ions: $Ca_3N_2$.

80.
a. silver(I) oxide or just silver oxide
b. correct
c. iron(III) oxide or ferric oxide
d. lead(IV) oxide or plumbic oxide
e. correct

82.
a. stannous chloride
b. ferrous oxide
c. stannic oxide
d. plumbous sulfide
e. cobaltic sulfide
f. chromous chloride

84.
a. iron(III) acetate, ferric acetate
b. bromine monofluoride
c. potassium peroxide
d. silicon tetrabromide
e. copper(II) permanganate, cupric permanganate
f. calcium chromate

86. (*a*); The correct name is aluminum sulfide.

88.
a. carbonate
b. chlorate
c. sulfate
d. phosphate
e. perchlorate
f. permanganate

90. RbCl; Alkali metals form a 1+ charge when bonding with a nonmetal to form an ionic compound. Since the ion contains 36 electrons and has a 1+ charge, this means it has one more proton as compared to electrons. The number of protons is therefore 37, identifying the ion as rubidium.

92.
a. $NaH_2PO_4$
b. $LiClO_4$
c. $Cu(HCO_3)_2$

Chapter 5:   Nomenclature

    d.    $KC_2H_3O_2$

    e.    $BaO_2$

    f.    $Cs_2SO_3$

94.

| Atom | G or L | Ion |
|---|---|---|
| K | L | $K^+$ |
| Cs | L | $Cs^+$ |
| Br | G | $Br^-$ |
| S | G | $S^{2-}$ |
| Se | G | $Se^{2-}$ |

96.

| Formula | Compound Name |
|---|---|
| $Co(NO_2)_2$ | cobalt(II) nitrite or cobaltous nitrite |
| $AsF_5$ | arsenic pentafluoride |
| LiCN | lithium cyanide |
| $K_2SO_3$ | potassium sulfite |
| $Li_3N$ | lithium nitride |
| $PbCrO_4$ | lead(II) chromate or plumbous chromate |

98.    (*b*) and (*d*); The symbols for the elements magnesium, aluminum, and xenon are *Mg*, Al, and Xe, respectively. Ga is expected to *lose* electrons to form ions in ionic compounds. The correct name for $TiO_2$ is *titanium(IV) oxide*.

# CUMULATIVE REVIEW

# Chapters 4 and 5

2. How many elements could you name? Although you certainly don't have to memorize all the elements, you should at least be able to give the symbol or name for the most common elements (listed in Table 4.3).

4. Dalton's atomic theory as presented in this text consists of five main postulates. Although Dalton's theory was exceptional scientific thinking for its time, some of the postulates have been modified as our scientific instruments and calculation methods have become increasingly more sophisticated. The main postulates of Dalton's theory are as follows: (1) Elements are made up of tiny particles called atoms; (2) all atoms of a given element are identical; (3) although all atoms of a given element are identical, these atoms are different from the atoms of all other elements; (4) atoms of one element can combine with atoms of another element to form a compound, and such a compound will always have the same relative numbers and types of atoms for its composition; and (5) atoms are merely rearranged into new groupings during an ordinary chemical reaction, and no atom is ever destroyed and no new atom is ever created during such a reaction.

6. The expression *nuclear* atom indicates that we view the atom as having a dense center of positive charge (called the nucleus) around which the electrons move through primarily empty space. Rutherford's experiment involved shooting a beam of particles at a thin sheet of metal foil. According to the then current "plum pudding" model of the atom, most of these positively-charged particles should have passed right through the foil. However, Rutherford detected that a significant number of particles effectively bounced off something and were deflected backwards to the source of particles, and that other particles were deflected from the foil at large angles. Rutherford realized that his observations could be explained if the atoms of the metal foil had a small, dense, positively-charged nucleus, with a significant amount of empty space between nuclei. The empty space between nuclei would allow most of the particles to pass through the atom. However, if a particle hit a nucleus head-on, it would be deflected backwards at the source. If a positively-charged particle passed near a positively-charged nucleus (but did not hit the nucleus head-on), then the particle would be deflected by the repulsive forces between the positive charges. Rutherford's experiment conclusively disproved the "plum pudding" model for the atom, which envisioned the atom as a uniform sphere of positive charge, with enough negatively-charged electrons scattered through the atom to balance out the positive charge.

8. Isotopes represent atoms of the same element which have different atomic masses. Isotopes are a result of the fact that atoms of a given element may have different numbers of neutrons in their nuclei. Isotopes have the same atomic number (number of protons in the nucleus) but have different mass numbers (total number of protons and neutrons in the nucleus). The different isotopes of an atom are indicated by symbolism of the form $^A_Z X$ in which $Z$ represents the atomic number, and $A$ the mass number, of element X. For example, $^{13}_{6}C$ represents a nuclide of carbon with atomic number 6 (6 protons in the nucleus) and mass number 13 (reflecting 6 protons plus 7 neutrons in the nucleus). The various isotopes of an element have identical chemical properties because the chemical properties of an atom are a function of the electrons in the atom (*not* the

Review: Chapter 4 and 5

nucleus). The physical properties of the isotopes of an element (and compounds containing those isotopes) may differ because of the difference in mass of the isotopes.

10. Most elements are too reactive to be found in nature in other than the combined form. Aside from the noble metals gold, silver, and platinum, the only other elements commonly found in nature in the uncombined state are some of the gaseous elements (such as O2, N2, He, Ar, etc.), and the solid nonmetals carbon and sulfur.

12. Ionic compounds typically are hard, crystalline solids with high melting and boiling points. Ionic substances like sodium chloride, when dissolved in water or when melted, conduct electrical currents: chemists have taken this evidence to mean that ionic substances consist of positively– and negatively–charged particles (ions). Although an ionic substance is made up of positively– and negatively–charged particles, there is no net electrical charge on a sample of such a substance because the total number of positive charges is balanced by an equal number of negative charges. An ionic compound could not possibly exist of just cations or just anions: there must be a balance of charge or the compound would be very unstable (like charges repel each other).

14. When naming ionic compounds, we name the positive ion (cation) first. For simple binary Type I ionic compounds, the ending –*ide* is added to the root name of the element that is the negative ion (anion). For example, for the Type I ionic compound formed between potassium and sulfur, $K_2S$, the name would be potassium sulfide: potassium is the cation, sulfur is the anion (with the suffix –*ide* added). Type II compounds are named by either of two systems, the "*ous–ic*" system (which is falling out of use), and the "Roman numeral" system which is preferred by most chemists. Type II compounds involve elements that form more than one stable ion. It is therefore necessary to specify *which* ion is present in a given compound. For example, iron forms two types of stable ion: $Fe^{2+}$ and $Fe^{3+}$. Iron can react with oxygen to form either of two stable oxides, FeO or $Fe_2O_3$, depending on which cation is involved. Under the Roman numeral naming system, FeO would be named iron(II) oxide to show that it contains $Fe^{2+}$ ions; $Fe_2O_3$ would be named iron(III) oxide to indicate that it contains $Fe^{3+}$ ions. The Roman numeral used in a name corresponds to the charge of the specific ion present in the compound. Under the less-favored "ous–ic" system, for an element that forms two stable ions, the ending –*ous* is used to indicate the lower-charged ion, whereas the ending –*ic* is used to indicate the higher-charged ion. FeO and $Fe_2O_3$ would thus be named ferr*ous* oxide and ferr*ic* oxide, respectively. The "ous–ic" system has fallen out of favor because it does not indicate the actual charge on the ion, but only that it is the lower or higher charged of the two. This can lead to confusion: for example $Fe^{2+}$ is called ferrous ion in this system, but $Cu^{2+}$ is called cupric ion (since there is also a $Cu^+$ stable ion).

16. A polyatomic ion is an ion containing more than one atom. Some common polyatomic ions you should be familiar with are listed in Table 5.4. Parentheses are used in writing formulas containing polyatomic ions to indicate unambiguously how many of the polyatomic ion are present in the formula, to make certain that there is no mistake as to what is meant by the formula. For example, consider the substance calcium phosphate. The correct formula for this substance is $Ca_3(PO_4)_2$, which indicates that three calcium ions are combined for every two phosphate ions (check the total number of positive and negative charges to see why this is so). If we did not write the parenthesis around the formula for the phosphate ion, that is, if we had written $Ca_3PO_{42}$, people reading this formula might think that there were 42 oxygen atoms present!

18. Acids, in general, are substances that produce protons ($H^+$ ions) when dissolved in water. For acids that do not contain oxygen, the prefix *hydro*– and the suffix –*ic* are used with the root name of the element present in the acid (for example: HCl, hydrochloric acid; $H_2S$, hydrosulfuric acid;

HF, hydrofluoric acid). The nomenclature of acids whose anions contain oxygen is more complicated. A series of prefixes and suffixes is used with the name of the non-oxygen atom in the anion of the acid: these prefixes and suffixes indicate the relative (not actual) number of oxygen atoms present in the anion. Most of the elements that form oxyanions form two such anions: for example, sulfur forms sulfite ion ($SO_3^{2-}$) and sulfate ion ($SO_4^{2-}$), and nitrogen forms nitrite ion ($NO_2^-$) and nitrate ion ($NO_3^-$). For an element that forms two oxyanions, the acid containing the anions will have the ending *–ous* if the anion is the *–ite* anion and the ending *–ic* if the anion is the *–ate* anion. For example, $HNO_2$ is nit*rous* acid and $HNO_3$ is nit*ric* acid; $H_2SO_3$ is sulf*urous* acid and $H_2SO_4$ is sulf*uric* acid. The halogen elements (Group 7) each form four oxyanions, and consequently, four oxyacids. The prefix *hypo–* is used for the oxyacid that contains fewer oxygen atoms than the *–ite* anion, and the prefix *per–* is used for the oxyacid that contains more oxygen atoms than the *–ate* anion. For example,

| Acid | Name | Anion | Anion name |
|---|---|---|---|
| HBrO | *hypo*brom*ous* acid | $BrO^-$ | *hypo*brom*ite* |
| $HBrO_2$ | brom*ous* acid | $BrO_2^-$ | brom*ite* |
| $HBrO_3$ | brom*ic* acid | $BrO_3^-$ | brom*ate* |
| $HBrO_4$ | *per*brom*ic* acid | $BrO_4^-$ | *per*brom*ate* |

20. How many elements in each family could you name? Elements in the same family have the same type of electronic configuration, and tend to undergo similar chemical reactions with other groups. For example, Li, Na, K, Rb, Cs all react with elemental chlorine gas, $Cl_2$, to form an ionic compound of general formula $M^+Cl^-$.

22. 
a. 8 electrons, 8 protons, 9 neutrons
b. 92 electrons, 92 protons, 143 neutrons
c. 17 electrons, 17 protons, 20 neutrons
d. 1 electrons, 1 protons, 2 neutrons
e. 2 electrons, 2 protons, 2 neutrons
f. 50 electrons, 50 protons, 69 neutrons
g. 54 electrons, 54 protons, 70 neutrons
h. 30 electrons, 30 protons, 34 neutrons

24. 
a. 12 protons, 10 electrons
b. 26 protons, 24 electrons
c. 26 protons, 23 electrons
d. 9 protons, 10 electrons
e. 28 protons, 26 electrons
f. 30 protons, 28 electrons
g. 27 protons, 24 electrons
h. 7 protons, 10 electrons
i. 16 protons, 18 electrons

Review: Chapter 4 and 5

      j.    37 protons, 36 electrons
      k.    34 protons, 36 electrons
      l.    19 protons, 18 electrons

26.    a.    CuI
      b.    $CoCl_2$
      c.    $Ag_2S$
      d.    $Hg_2Br_2$
      e.    HgO
      f.    $Cr_2S_3$
      g.    $PbO_2$
      h.    $K_3N$
      i.    $SnF_2$
      j.    $Fe_2O_3$

28.    a.    $CO_3^{2-}$, carbonate ion
      b.    $MnO_4^-$, permanganate ion
      c.    $NO_3^-$, nitrate ion
      d.    $HSO_4^-$, hydrogen sulfate (bisulfate) ion
      e.    $C_2H_3O_2^-$, acetate ion
      f.    $CrO_4^{2-}$, chromate ion
      g.    $OH^-$, hydroxide ion
      h.    $ClO_2^-$, chlorite ion
      i.    $HCO_3^-$, hydrogen carbonate (bicarbonate) ion
      j.    $HPO_4^{2-}$, hydrogen phosphate ion

30.    a.    xenon dioxide
      b.    iodine pentachloride
      c.    phosphorus trichloride
      d.    carbon monoxide
      e.    oxygen difluoride
      f.    diphosphorus pentoxide
      g.    arsenic triiodide
      h.    sulfur trioxide

# CHAPTER 6

# Chemical Reactions: An Introduction

2. Most of these products contain a peroxide, which decomposes releasing oxygen gas.

4. Bubbling takes place as the hydrogen peroxide chemically decomposes into water and oxygen gas.

6. The appearance of the black color actually signals the breakdown of starches and sugars in the bread to elemental carbon. You may also see steam coming from the bread (water produced by the breakdown of the carbohydrates).

8. a. HCl, $O_2$; The reactants are on the left side of the arrow.
   b. $H_2O$, $Cl_2$; The products are on the right side of the arrow.

10. Balancing an equation ensures that no atoms are created or destroyed during the reaction. The total mass after the reaction must be the same as the total mass before the reaction.

12. liquid

14. $H_2O_2 \rightarrow H_2O + O_2$

16. $N_2H_4(l) \rightarrow N_2(g) + H_2(g)$

18. $Ag_2O(s) \rightarrow Ag(s) + O_2(g)$

20. $CaCO_3(s) + HCl(aq) \rightarrow CaCl_2(aq) + H_2O(l) + CO_2(g)$

22. $SiO_2(s) + C(s) \rightarrow Si(s) + CO(g)$

24. $Zn(s) + HCl(aq) \rightarrow H_2(g) + ZnCl_2(aq)$

26. $SO_2(g) + H_2O(l) \rightarrow H_2SO_3(aq)$
    $SO_3(g) + H_2O(l) \rightarrow H_2SO_4(aq)$

28. $NO(g) + O_3(g) \rightarrow NO_2(g) + O_2(g)$

30. $P_4(s) + O_2(g) \rightarrow P_2O_5(s)$

32. $Xe(g) + F_2(g) \rightarrow XeF_4(s)$

34. $NH_3(g) + O_2(g) \rightarrow HNO_3(aq) + H_2O(l)$

Chapter 6:   Chemical Reactions:   An Introduction

36. We cannot change the identities or formulas of the reactants or products in a chemical equation when balancing the equation. The proposed equation has incorrectly changed one of the products from water to hydrogen gas. (Use molecular-level drawings to support your answer.)

38. $2K(s) + 2H_2O(l) \rightarrow H_2(g) + 2KOH(aq)$

40. 
a. $Na_2SO_4(aq) + CaCl_2(aq) \rightarrow CaSO_4(s) + 2NaCl(aq)$
b. $3Fe(s) + 4H_2O(g) \rightarrow Fe_3O_4(s) + 4H_2(g)$
c. $Ca(OH)_2(aq) + 2HCl(aq) \rightarrow CaCl_2(aq) + 2H_2O(l)$
d. $Br_2(g) + 2H_2O(l) + SO_2(g) \rightarrow 2HBr(aq) + H_2SO_4(aq)$
e. $3NaOH(s) + H_3PO_4(aq) \rightarrow Na_3PO_4(aq) + 3H_2O(l)$
f. $2NaNO_3(s) \rightarrow 2NaNO_2(s) + O_2(g)$
g. $2Na_2O_2(s) + 2H_2O(l) \rightarrow 4NaOH(aq) + O_2(g)$
h. $4Si(s) + S_8(s) \rightarrow 2Si_2S_4(s)$

42. 
a. $4NaCl(s) + 2SO_2(g) + 2H_2O(g) + O_2(g) \rightarrow 2Na_2SO_4(s) + 4HCl(g)$
b. $3Br_2(l) + I_2(s) \rightarrow 2IBr_3(s)$
c. $Ca(s) + 2H_2O(g) \rightarrow Ca(OH)_2(aq) + H_2(g)$
d. $2BF_3(g) + 3H_2O(g) \rightarrow B_2O_3(s) + 6HF(g)$
e. $SO_2(g) + 2Cl_2(g) \rightarrow SOCl_2(l) + Cl_2O(g)$
f. $Li_2O(s) + H_2O(l) \rightarrow 2LiOH(aq)$
g. $Mg(s) + CuO(s) \rightarrow MgO(s) + Cu(l)$
h. $Fe_3O_4(s) + 4H_2(g) \rightarrow 3Fe(l) + 4H_2O(g)$

44. 
a. $Ba(NO_3)_2(aq) + Na_2CrO_4(aq) \rightarrow BaCrO_4(s) + 2NaNO_3(aq)$
b. $PbCl_2(aq) + K_2SO_4(aq) \rightarrow PbSO_4(s) + 2KCl(aq)$
c. $C_2H_5OH(l) + 3O_2(g) \rightarrow 2CO_2(g) + 3H_2O(l)$
d. $CaC_2(s) + 2H_2O(l) \rightarrow Ca(OH)_2(s) + C_2H_2(g)$
e. $Sr(s) + 2HNO_3(aq) \rightarrow Sr(NO_3)_2(aq) + H_2(g)$
f. $BaO_2(s) + H_2SO_4(aq) \rightarrow BaSO_4(s) + H_2O_2(aq)$
g. $2AsI_3(s) \rightarrow 2As(s) + 3I_2(s)$
h. $2CuSO_4(aq) + 4KI(s) \rightarrow 2CuI(s) + I_2(s) + 2K_2SO_4(aq)$

46. (a)

48.
$CaC_2(s) + 2H_2O(l) \rightarrow C_2H_2(g) + Ca(OH)_2(aq)$
$2C_2H_2(g) + 5O_2(g) \rightarrow 4CO_2(g) + 2H_2O(g)$

50. $2Al_2O_3(s) + 3C(s) \rightarrow 4Al(s) + 3CO_2(g)$

Chapter 6:    Chemical Reactions:    An Introduction

52. True. Coefficients can be fractions when balancing a chemical equation because the coefficients represent a ratio of the moles needed for the reaction to occur. As a result, moles can be fractions because it represents an amount. The key is to make sure the atoms are conserved from reactants to products. Take note that the accepted convention is that the "best" balanced equation is the one with the smallest integers (although not required).

54. $BaO_2(s) + H_2O(l) \rightarrow BaO(s) + H_2O_2(aq)$

56. $2KClO_3(s) \rightarrow 2KCl(s) + 3O_2(g)$

58. $2C_7H_{14} + 21O_2 \rightarrow 14CO_2 + 14H_2O$ (balanced in standard form)

60. $3LiAlH_4(s) + AlCl_3(s) \rightarrow 4AlH_3(s) + 3LiCl(s)$

62. $Fe(s) + S(s) \rightarrow FeS(s)$

64. $K_2CrO_4(aq) + BaCl_2(aq) \rightarrow BaCrO_4(s) + 2KCl(aq)$

66. $2NaCl(aq) + 2H_2O(l) \rightarrow 2NaOH(aq) + H_2(g) + Cl_2(g)$

$2NaBr(aq) + 2H_2O(l) \rightarrow 2NaOH(aq) + H_2(g) + Br_2(g)$

$2NaI(aq) + 2H_2O(l) \rightarrow 2NaOH(aq) + H_2(g) + I_2(g)$

68. (e); Subscripts cannot be changed to balance an equation. If the subscripts are changed, then the identity of at least one of the compounds will change.

70. $CuO(s) + H_2SO_4(aq) \rightarrow CuSO_4(aq) + H_2O(l)$

72. (a)

74. $2CuSO_4(aq) + 4KI(s) \rightarrow 2CuI(s) + I_2(s) + 2K_2SO_4(aq)$

76. Answers will vary but the following balanced equation should be reported:

$4NH_3(g) + 5O_2(g) \rightarrow 4NO(g) + 6H_2O(g)$

78. $4Fe(s) + 3O_2(g) \rightarrow 2Fe_2O_3(s)$

$2PbO_2(s) \rightarrow 2PbO(s) + O_2(g)$

$2H_2O_2(l) \rightarrow O_2(g) + 2H_2O(l)$

# CHAPTER 7

# Reactions in Aqueous Solution

2. Driving forces are types of *changes* in a system that pull a reaction in the *direction of product formation*; driving forces discussed in Chapter 7 include: formation of a *solid*, formation of *water*, formation of a *gas*, and transfer of electrons.

4. A reactant in aqueous solution is indicated with (*aq*). Formation of a solid is indicated with (*s*)

6. Because each formula unit of $K_3PO_4$ contains three potassium ions for each phosphate ion, that ratio will be preserved in the solution when $K_3PO_4$ is dissolved in water.

8. Chemists know that a solution contains independent ions because such a solution will readily allow an electrical current to pass through it. The simplest experiment that demonstrates this uses the sort of light–bulb conductivity apparatus described in the text: if the light bulb glows strongly, then the solution must contain a lot of ions to be conducting the electricity well.

10. (*a*); The precipitate $BaSO_4$ will form.

12. a. soluble (Rule 1: most nitrate salts are soluble.)
    b. soluble (Rule 2: most salts of $K^+$ are soluble.)
    c. insoluble (Rule 4: most sulfate salts are soluble with $PbSO_4$ as an exception.)
    d. insoluble (Rule 5: most hydroxide compounds are insoluble.)
    e. soluble (Rule 2: most salts of $K^+$ are soluble.)
    f. insoluble (Rule 3: most chloride salts are soluble with $Hg_2Cl_2$ as an exception.)
    g. soluble (Rule 2: most salts of $NH_4^+$ are soluble.)
    h. insoluble (Rule 6: most sulfide salts are insoluble.)

14. a. Rule 5: Most hydroxide compounds are insoluble.
    b. Rule 6: Most carbonate salts are insoluble.
    c. Rule 6: Most phosphate salts are insoluble.
    d. Rule 3: Exception to the rule for chloride salts.
    e. Rule 4: Exception to the rule for sulfate salts.

16. A precipitate will form in all four beakers. In beaker 1, the precipitate $PbCl_2$ will form (Rule 3 – exception for chlorides). In beaker 2, the precipitate $Pb(OH)_2$ will form (Rule 5 – most hydroxides are only slightly soluble). In beaker 3, the precipitate $Pb_3(PO_4)_2$ will form (Rule 6 – most phosphates are only slightly soluble). In beaker 4, the precipitate $PbSO_4$ will form (Rule 4 – exception for sulfates).

Chapter 7:  Reactions in Aqueous Solution

18. The formulas of the precipitates are in boldface type.

   a. Rule 6: Most carbonate salts are insoluble.

   $Na_2CO_3(aq) + CuSO_4(aq) \rightarrow Na_2SO_4(aq) + \mathbf{CuCO_3(s)}$

   b. Rule 3: Exception for chloride salts.

   $HCl(aq) + AgC_2H_3O_2(aq) \rightarrow HC_2H_3O_2(aq) + \mathbf{AgCl(s)}$

   c. No precipitate

   d. Rule 6: Most sulfide salts are insoluble.

   $3(NH_4)_2S(aq) + 2FeCl_3(aq) \rightarrow 6NH_4Cl(aq) + \mathbf{Fe_2S_3(s)}$

   e. Rule 4: Exception for sulfate salts

   $H_2SO_4(aq) + Pb(NO_3)_2(aq) \rightarrow 2HNO_3(aq) + \mathbf{PbSO_4(s)}$

   f. Rule 6: Most phosphate salts are insoluble.

   $2K_3PO_4(aq) + 3CaCl_2(aq) \rightarrow 6KCl(aq) + \mathbf{Ca_3(PO_4)_2(s)}$

20. Hint: when balancing equations involving polyatomic ions, especially in precipitation reactions, balance the polyatomic ions as a *unit*, not in terms of the atoms the polyatomic ions contain (e.g., treat nitrate ion, $NO_3^-$ as a single entity, not as one nitrogen and three oxygen atoms). When finished balancing, however, be sure to count the individual number of atoms of each type on each side of the equation.

   a. $CaCl_2(aq) + AgNO_3(aq) \rightarrow Ca(NO_3)_2(aq) + AgCl(s)$

   balance chlorine: $CaCl_2(aq) + AgNO_3(aq) \rightarrow Ca(NO_3)_2(aq) + \mathbf{2}AgCl(s)$

   balance silver: $CaCl_2(aq) + \mathbf{2}AgNO_3(aq) \rightarrow Ca(NO_3)_2(aq) + 2AgCl(s)$

   balanced equation: $CaCl_2(aq) + 2AgNO_3(aq) \rightarrow Ca(NO_3)_2(aq) + 2AgCl(s)$

   b. $AgNO_3(aq) + K_2CrO_4(aq) \rightarrow Ag_2CrO_4(s) + KNO_3(aq)$

   balance silver: $\mathbf{2}AgNO_3(aq) + K_2CrO_4(aq) \rightarrow Ag_2CrO_4(s) + KNO_3(aq)$

   balance nitrate ion: $2AgNO_3(aq) + K_2CrO_4(aq) \rightarrow Ag_2CrO_4(s) + \mathbf{2}KNO_3(aq)$

   balanced equation: $2AgNO_3(aq) + K_2CrO_4(aq) \rightarrow Ag_2CrO_4(s) + 2KNO_3(aq)$

   c. $BaCl_2(aq) + K_2SO_4(aq) \rightarrow BaSO_4(s) + KCl(aq)$

   balance potassium: $BaCl_2(aq) + K_2SO_4(aq) \rightarrow BaSO_4(s) + \mathbf{2}KCl(aq)$

   balanced equation: $BaCl_2(aq) + K_2SO_4(aq) \rightarrow BaSO_4(s) + 2KCl(aq)$

22. The balanced equation is:

   $Zn(NO_3)_2(aq) + 2KOH(aq) \rightarrow Zn(OH)_2(s) + 2KNO_3(aq)$

24. (e)

## Chapter 7: Reactions in Aqueous Solution

26. Molecular: $NiCl_2(aq) + Na_2S(aq) \rightarrow 2NaCl(aq) + NiS(s)$

    Complete Ionic: $Ni^{2+}(aq) + 2Cl^-(aq) + 2Na^+(aq) + S^{2-}(aq) \rightarrow 2Na^+(aq) + 2Cl^-(aq) + NiS(s)$

    Net Ionic: $Ni^{2+}(aq) + S^{2-}(aq) \rightarrow NiS(s)$

28. $Ag^+(aq) + Cl^-(aq) \rightarrow AgCl(s)$

    $Pb^{2+}(aq) + 2Cl^-(aq) \rightarrow PbCl_2(s)$

    $Hg_2^{2+}(aq) + 2Cl^-(aq) \rightarrow Hg_2Cl_2(s)$

30. $Co^{2+}(aq) + S^{2-}(aq) \rightarrow CoS(s)$

    $2Co^{3+}(aq) + 3S^{2-}(aq) \rightarrow Co_2S_3(s)$

    $Fe^{2+}(aq) + S^{2-}(aq) \rightarrow FeS(s)$

    $2Fe^{3+}(aq) + 3S^{2-}(aq) \rightarrow Fe_2S_3(s)$

32. Strong bases fully produce hydroxide ions when dissolved in water. The strong bases are also strong electrolytes. Strong electrolytes dissociate completely in water.

34. acids: $HCl, H_2SO_4, HNO_3, HClO_4, HBr$

    bases: $NaOH, KOH, RbOH, CsOH$

36. A salt is the ionic product remaining in solution when an acid neutralizes a base. For example, in the reaction $HCl(aq) + NaOH(aq) \rightarrow NaCl(aq) + H_2O(l)$ sodium chloride is the salt produced by the neutralization reaction.

38. $HBr(aq) \rightarrow H^+(aq) + Br^-(aq)$

    $HClO_4(aq) \rightarrow H^+(aq) + ClO_4^-(aq)$

40. In general, the salt formed in an aqueous acid–base reaction consists of the *positive ion of the base* involved in the reaction, combined with the *negative ion of the acid*. The hydrogen ion of the strong acid combines with the hydroxide ion of the strong base to produce water, which is the other product of the acid–base reactions.

    a. $H_2SO_4(aq) + 2KOH(aq) \rightarrow K_2SO_4(aq) + 2H_2O(l)$

    b. $HNO_3(aq) + NaOH(aq) \rightarrow NaNO_3(aq) + H_2O(l)$

    c. $2HCl(aq) + Ca(OH)_2(aq) \rightarrow CaCl_2(aq) + 2H_2O(l)$

    d. $2HClO_4(aq) + Ba(OH)_2(aq) \rightarrow Ba(ClO_4)_2(aq) + 2H_2O(l)$

42. Answer depends on student choice of example: $Na(s) + Cl_2(g) \rightarrow 2NaCl(s)$ is an example.

44. The aluminum atoms lose 3 electrons to become $Al^{3+}$ ions. $Fe^{3+}$ ions gain 3 electrons to become Fe atoms.

46. Each magnesium atom would lose two electrons. Each oxygen atom would gain two electrons (so the $O_2$ molecule would gain four electrons). Two magnesium atoms would be required to react with each oxygen, $O_2$, molecule. Magnesium ions are charged 2+, oxide ions are charged 2–.

48. $Na_2O_2$ is made up of $Na^+$ ions and $O_2^{2-}$ ions. Sodium atoms each lose one electron and oxygen atoms gain two electrons ($O_2$ gains two electrons).

50. a. $P_4(s) + O_2(g) \rightarrow P_4O_{10}(s)$

    balance oxygen: $P_4(s) + 5O_2(g) \rightarrow P_4O_{10}(s)$

    balanced equation: $P_4(s) + 5O_2(g) \rightarrow P_4O_{10}(s)$

    b. $MgO(s) + C(s) \rightarrow Mg(s) + CO(g)$

    This equation is already balanced.

    c. $Sr(s) + H_2O(l) \rightarrow Sr(OH)_2(aq) + H_2(g)$

    balance oxygen: $Sr(s) + 2H_2O(l) \rightarrow Sr(OH)_2(aq) + H_2(g)$

    balanced equation: $Sr(s) + 2H_2O(l) \rightarrow Sr(OH)_2(aq) + H_2(g)$

    d. $Co(s) + HCl(aq) \rightarrow CoCl_2(aq) + H_2(g)$

    balance hydrogen: $Co(s) + 2HCl(aq) \rightarrow CoCl_2(aq) + H_2(g)$

    balanced equation: $Co(s) + 2HCl(aq) \rightarrow CoCl_2(aq) + H_2(g)$

52. A reaction must be an oxidation–reduction reaction if any of the oxidation numbers of the atoms in the equation change. Aluminum changes oxidation state from 0 in Al to +3 (oxidation) in $Al_2O_3$ and $AlCl_3$; nitrogen changes oxidation state from –3 in $NH_4^+$ to +2 in NO (oxidation); chlorine changes oxidation state from +7 in $ClO_4^-$ to –1 in $AlCl_3$ (reduction)

54. For each reaction, the type of reaction is first identified, followed by some of the reasoning that leads to this choice (there may be more than one way in which you can recognize a particular type of reaction).

    a. oxidation–reduction (Oxygen changes from the combined state to the elemental state.)

    b. oxidation–reduction (Zinc changes from the elemental to the combined state; hydrogen changes from the combined to the elemental state.)

    c. acid–base ($H_2SO_4$ is a strong acid and NaOH is a strong base; water and a salt are formed.)

    d. acid–base, precipitation ($H_2SO_4$ is a strong acid, and $Ba(OH)_2$ is a base; water and a salt are formed; an insoluble product forms.)

    e. precipitation (From the Solubility Rules of Table 7.1, AgCl is only slightly soluble.)

    f. precipitation (From the Solubility Rules of Table 7.1, $Cu(OH)_2$ is only slightly soluble.)

    g. oxidation–reduction (Chlorine and fluorine change from the elemental to the combined state.)

    h. oxidation–reduction (Oxygen changes from the elemental to the combined state.)

    i. acid–base ($HNO_3$ is a strong acid and $Ca(OH)_2$ is a strong base; a salt and water are formed.)

56. oxidation–reduction

## Chapter 7: Reactions in Aqueous Solution

**58.** A decomposition reaction is one in which a given compound is broken down into simpler compounds or constituent elements. The reactions

$$CaCO_3(s) \rightarrow CaO(s) + CO_2(g)$$

$$2HgO(s) \rightarrow 2Hg(l) + O_2(g)$$

both represent decomposition reactions. Such reactions often (but not necessarily always) may be classified in other ways. For example, the reaction of $HgO(s)$ is also an oxidation–reduction reaction.

**60.** Compounds like those in this problem, containing only carbon and hydrogen, are called *hydrocarbons*. When a hydrocarbon is reacted with oxygen ($O_2$), the hydrocarbon is almost always converted to carbon dioxide and water vapor. Because water molecules contain an odd number of oxygen atoms, and $O_2$ contains an even number of oxygen atoms, it is often difficult to balance such equations. For this reason, it is simpler to balance the equation using fractional coefficients if necessary, and then to multiply by a factor that will give whole number coefficients for the final balanced equation.

a. $CH_4(g) + 2O_2(g) \rightarrow CO_2(g) + 2H_2O(g)$

balance carbon: $CH_4(g) + O_2(g) \rightarrow CO_2(g) + H_2O(g)$

balance hydrogen: $CH_4(g) + O_2(g) \rightarrow CO_2(g) + \mathbf{2}H_2O(g)$

balance oxygen: $CH_4(g) + \mathbf{2}O_2(g) \rightarrow CO_2(g) + 2H_2O(g)$

balanced equation: $CH_4(g) + 2O_2(g) \rightarrow CO_2(g) + 2H_2O(g)$

b. $2C_2H_2(g) + 5O_2(g) \rightarrow 4CO_2(g) + 2H_2O(l)$

balance carbon: $C_2H_2(g) + O_2(g) \rightarrow \mathbf{2}CO_2(g) + H_2O(l)$

balance hydrogen: $C_2H_2(g) + O_2(g) \rightarrow 2CO_2(g) + H_2O(l)$

balance oxygen: $C_2H_2(g) + \frac{5}{2}O_2(g) \rightarrow 2CO_2(g) + H_2O(l)$

balanced equation: $2C_2H_2(g) + 5O_2(g) \rightarrow 4CO_2(g) + 2H_2O(l)$

c. $C_{10}H_8(s) + 12O_2(g) \rightarrow 10CO_2(g) + 4H_2O(l)$

balance carbon: $C_{10}H_8(s) + O_2(g) \rightarrow \mathbf{10}CO_2(g) + H_2O(l)$

balance hydrogen: $C_{10}H_8(s) + O_2(g) \rightarrow 10CO_2(g) + \mathbf{4}H_2O(l)$

balance oxygen: $C_{10}H_8(s) + 12O_2(g) \rightarrow 10CO_2(g) + 4H_2O(l)$

balanced equation: $C_{10}H_8(s) + 12O_2(g) \rightarrow 10CO_2(g) + 4H_2O(l)$

**62.** A reaction in which small molecules or atoms combine to make a larger molecule is called a *synthesis* reaction. An example would be the synthesis of sodium chloride from the elements

$$2Na(s) + Cl_2(g) \rightarrow 2NaCl(s).$$

A reaction in which a molecule is broken down into simpler molecules or atoms is called a *decomposition* reaction. An example would be the decomposition of sodium hydrogen carbonate when heated.

$$2NaHCO_3(s) \rightarrow Na_2CO_3(s) + CO_2(g) + H_2O(g).$$

Chapter 7: Reactions in Aqueous Solution

Specific examples will depend on the students' input.

64.  a.  $8Fe(s) + S_8(s) \rightarrow 8FeS(s)$
     b.  $4Co(s) + 3O_2(g) \rightarrow 2Co_2O_3(s)$
     c.  $Cl_2O_7(g) + H_2O(l) \rightarrow 2HClO_4(aq)$

66.  a.  $2Al(s) + 3Br_2(l) \rightarrow 2AlBr_3(s)$
     b.  $Zn(s) + 2HClO_4(aq) \rightarrow Zn(ClO_4)_2(aq) + H_2(g)$
     c.  $3Na(s) + P(s) \rightarrow Na_3P(s)$
     d.  $CH_4(g) + 4Cl_2(g) \rightarrow CCl_4(l) + 4HCl(g)$
     e.  $Cu(s) + 2AgNO_3(aq) \rightarrow Cu(NO_3)_2(aq) + 2Ag(s)$

68.  (c) and (d); (a) and (b) will form insoluble compounds with $Ag^+$. For (e), a compound will not form between $Na^+$ and $Ag^+$.

70.  The formulas of the salts are indicated in boldface type.
     a.  $HNO_3(aq) + KOH(aq) \rightarrow H_2O(l) + \mathbf{KNO_3}(aq)$
     b.  $H_2SO_4(aq) + Ba(OH)_2(aq) \rightarrow 2H_2O(l) + \mathbf{BaSO_4}(s)$
     c.  $HClO_4(aq) + NaOH(aq) \rightarrow H_2O(l) + \mathbf{NaClO_4}(aq)$
     d.  $2HCl(aq) + Ca(OH)_2(aq) \rightarrow 2H_2O(l) + \mathbf{CaCl_2}(aq)$

72.  (a), (b), and (c). There are no precipitates formed in (d); $Ca(NO_3)_2$ is soluble.

74.  a.  $H_2SO_4(aq) + 2NaOH(aq) \rightarrow 2H_2O(l) + Na_2SO_4(aq)$
     b.  $HNO_3(aq) + RbOH(aq) \rightarrow H_2O(l) + RbNO_3(aq)$
     c.  $HClO_4(aq) + KOH(aq) \rightarrow 2H_2O(l) + KClO_4(aq)$
     d.  $HCl(aq) + KOH(aq) \rightarrow H_2O(l) + KCl(aq)$

76.  Molecular: $3(NH_4)_2S(aq) + 2FeCl_3(aq) \rightarrow Fe_2S_3(s) + 6NH_4Cl(aq)$

Complete Ionic: $6NH_4^+(aq) + 3S^{2-}(aq) + 2Fe^{3+}(aq) + 6Cl^-(aq) \rightarrow Fe_2S_3(s) + 6NH_4^+(aq) + 6Cl^-(aq)$

Net Ionic: $3S^{2-}(aq) + 2Fe^{3+}(aq) \rightarrow Fe_2S_3(s)$

78.  Aluminum atoms lose 3 electrons to become $Al^{3+}$ ions. Iodine atoms gain 1 electron each to become $I^-$ ions.

80.  a.  $Na + O_2 \rightarrow Na_2O_2$

     Balance sodium: $\mathbf{2}Na + O_2 \rightarrow Na_2O_2$

     Balanced equation: $2Na(s) + O_2(g) \rightarrow Na_2O_2(s)$

     b.  $Fe(s) + H_2SO_4(aq) \rightarrow FeSO_4(aq) + H_2(g)$

     Equation is already balanced!

## Chapter 7: Reactions in Aqueous Solution

c. $Al_2O_3 \rightarrow Al + O_2$

  Balance oxygen: $2Al_2O_3 \rightarrow Al + 3O_2$

  Balance aluminum: $2Al_2O_3 \rightarrow 4Al + 3O_2$

  Balanced equation: $2Al_2O_3(s) \rightarrow 4Al(s) + 3O_2(g)$

d. $Fe + Br_2 \rightarrow FeBr_3$

  Balance bromine: $Fe + 3Br_2 \rightarrow 2FeBr_3$

  Balance iron: $2Fe + 3Br_2 \rightarrow 2FeBr_3$

  Balanced equation: $2Fe(s) + 3Br_2(l) \rightarrow 2FeBr_3(s)$

e. $Zn + HNO_3 \rightarrow Zn(NO_3)_2 + H_2$

  Balance nitrate ions: $Zn + 2HNO_3 \rightarrow Zn(NO_3)_2 + H_2$

  Balanced equation: $Zn(s) + 2HNO_3(aq) \rightarrow Zn(NO_3)_2(aq) + H_2(g)$

82. (a), (b), and (c); All three statements are true.

84. 
  a. $2NaHCO_3(s) \rightarrow Na_2CO_3(s) + H_2O(g) + CO_2(g)$
  b. $2NaClO_3(s) \rightarrow 2NaCl(s) + 3O_2(g)$
  c. $2HgO(s) \rightarrow 2Hg(l) + O_2(g)$
  d. $C_{12}H_{22}O_{11}(s) \rightarrow 12C(s) + 11H_2O(g)$
  e. $2H_2O_2(l) \rightarrow 2H_2O(l) + O_2(g)$

86.

|  | $CaCl_2$ | $Pb(NO_3)_2$ | $(NH_4)_3PO_4$ |
|---|---|---|---|
| $Na_2CO_3$ | $CaCO_3$ | $PbCO_3$ | No |
| $AgNO_3$ | $AgCl$ | No | $Ag_3PO_4$ |
| $K_2SO_4$ | $CaSO_4$ | $PbSO_4$ | No |

The balanced equations are (moving across the table by row):

$Na_2CO_3(aq) + CaCl_2(aq) \rightarrow CaCO_3(s) + 2NaCl(aq)$

$Na_2CO_3(aq) + Pb(NO_3)_2(aq) \rightarrow PbCO_3(s) + 2NaNO_3(aq)$

$3Na_2CO_3(aq) + 2(NH_4)_3PO_4(aq) \rightarrow 3(NH_4)_2CO_3(aq) + 2Na_3PO_4(aq)$; no precipitate formed

$2AgNO_3(aq) + CaCl_2(aq) \rightarrow 2AgCl(s) + Ca(NO_3)_2(aq)$

Chapter 7: Reactions in Aqueous Solution

$AgNO_3(aq) + Pb(NO_3)_2(aq) \rightarrow AgNO_3(aq) + Pb(NO_3)_2(aq)$; no precipitate formed

$3AgNO_3(aq) + (NH_4)_3PO_4(aq) \rightarrow \mathbf{Ag_3PO_4}(s) + 3NH_4NO_3(aq)$

$K_2SO_4(aq) + CaCl_2(aq) \rightarrow \mathbf{CaSO_4}(s) + 2KCl(aq)$

$K_2SO_4(aq) + Pb(NO_3)_2(aq) \rightarrow \mathbf{PbSO_4}(s) + 2KNO_3(aq)$

$3K_2SO_4(aq) + 2(NH_4)_3PO_4(aq) \rightarrow 3(NH_4)_2SO_4(aq) + 2K_3PO_4(aq)$; no precipitate formed

88.     a.     one
       b.     one
       c.     two
       d.     two
       e.     three

90.     False. The balanced molecular equation is: $Ba(OH)_2(aq) + H_2SO_4(aq) \rightarrow BaSO_4(s) + 2H_2O(l)$. The complete ionic equation is: $Ba^{2+}(aq) + 2OH^-(aq) + 2H^+(aq) + SO_4^{2-}(aq) \rightarrow BaSO_4(s) + 2H_2O(l)$. The net ionic equation includes all species that take part in the chemical reaction. The $OH^-$ and $H^+$ ions form water so they are also included in the net ionic equation. Thus the complete ionic equation and net ionic equation are the same.

92.     $3Na_2CrO_4(aq) + 2AlBr_3(aq) \rightarrow Al_2(CrO_4)_3(s) + 6NaBr(aq)$

94.     $PbCl_2$; $PbSO_4$; $Pb_3(PO_4)_2$; $AgCl$; $Ag_3PO_4$

96.     $PbSO_4$; $AgCl$; none

# CUMULATIVE REVIEW

# Chapters 6 and 7

2. A chemical equation indicates the substances necessary for a chemical reaction to take place, as well as what is produced by that chemical reaction. The substances to the left of the arrow in a chemical equation are called the reactants; those to the right of the arrow are referred to as the products. In addition, if a chemical equation has been balanced, then the equation indicates the relative proportions in which the reactant molecules combine to form the product molecules.

4. It is *never* permissible to change the subscripts of a formula when balancing a chemical equation: changing the subscripts changes the *identity* of a substance from one chemical to another. For example, consider the unbalanced chemical equation

   $$H_2(g) + O_2(g) \rightarrow H_2O(l).$$

   If you changed the *formula* of the product from $H_2O(l)$ to $H_2O_2(l)$, the equation would appear to be "balanced". However, $H_2O$ is water, whereas $H_2O_2$ is hydrogen peroxide–a completely different chemical substance (which is not prepared by reaction of the elements hydrogen and oxygen).

   When we balance a chemical equation, it is permitted only to adjust the *coefficients* of a formula, so changing a coefficient merely changes the number of molecules of a substance being used in the reaction, without changing the identity of the substance. For the example above, we can balance the equation by putting coefficients of 2 in front of the formulas of $H_2$ and $H_2O$: these coefficients do not change the nature of what is reacting and what product is formed.

   $$2H_2(g) + O_2(g) \rightarrow 2H_2O(l)$$

6. A precipitation reaction is one in which a *solid* forms when the reactants are combined: the solid is called a precipitate. The driving force in such a reaction is the formation of the solid, thus *removing ions* from the solution. There are many examples of such precipitation reactions: consult the solubility rules in Table 7.1 if you need help. One example would be to combine barium nitrate and sodium carbonate solutions: a precipitate of barium carbonate would form.

   The molecular equation for this reaction is:

   $$Ba(NO_3)_2(aq) + Na_2CO_3(aq) \rightarrow BaCO_3(s) + 2NaNO_3(aq)$$

   The net ionic equation for this reaction is:

   $$Ba^{2+}(aq) + CO_3^{2-}(aq) \rightarrow BaCO_3(s)$$

8. In summary, nearly all compounds containing the nitrate, sodium, potassium, and ammonium ions are soluble in water. Most salts containing the chloride and sulfate ions are soluble in water, with specific exceptions (see Table 7.1 for these exceptions). Most compounds containing the hydroxide, sulfide, carbonate, and phosphate ions are not soluble in water, unless the compound also contains one of the cations mentioned above ($Na^+$, $K^+$, $NH_4^+$).

   The solubility rules are phrased as if you had a sample of a given solute and wanted to see if you could dissolve it in water. These rules can also be applied, however, to predict the identity of the solid produced in a precipitation reaction: a given combination of ions will not be soluble in water whether you take a pure compound out of a reagent bottle or if you generate the insoluble combination of ions during a chemical reaction. For example, the solubility rules say that $BaSO_4$ is not soluble in water. This means not only that a pure sample of $BaSO_4$ taken from a reagent bottle will not dissolve in water, but also that if $Ba^{2+}$ ion and $SO_4^{2-}$ ion end up together in the same solution, they will precipitate as $BaSO_4$. If we were to combine barium chloride and sulfuric acid solutions

$$BaCl_2(aq) + H_2SO_4(aq) \rightarrow BaSO_4(s) + 2HCl(aq)$$

then, because barium sulfate is not soluble in water, a precipitate of $BaSO_4(s)$ would form. Because a precipitate of $BaSO_4(s)$ would form no matter what barium compound or what sulfate compound were mixed, we can write the net ionic equation for the reaction as

$$Ba^{2+}(aq) + SO_4^{2-}(aq) \rightarrow BaSO_4(s).$$

Thus if, for example, barium nitrate solution were combined with sodium sulfate solution, a precipitate of $BaSO_4$ would form. Barium sulfate is insoluble in water regardless of its source.

10. Acids (such as the citric acid found in citrus fruits and the acetic acid found in vinegar) were first noted primarily because of their sour taste. The first bases noted were characterized by their bitter taste and slippery feel on the skin. Acids and bases chemically react with (neutralize) each other forming water: the net ionic equation is

$$H^+(aq) + OH^-(aq) \rightarrow H_2O(l).$$

The *strong* acids and bases fully ionize when they dissolve in water: because these substances fully ionize, they are strong electrolytes. The common strong acids are HCl(hydrochloric), $HNO_3$(nitric), $H_2SO_4$(sulfuric), and $HClO_4$(perchloric). The most common strong bases are the alkali metal hydroxides, particularly NaOH(sodium hydroxide) and KOH(potassium hydroxide).

12. Oxidation–reduction reactions are electron-transfer reactions. Oxidation represents a loss of electrons by an atom, molecule, or ion, whereas reduction is the gain of electrons by such a species. Because an oxidation–reduction process represents the transfer of electrons between species, you can't have one without the other also taking place: the electrons lost by one species must be gained by some other species. An example of a simple oxidation–reduction reaction between a metal and a nonmetal could be the following

$$Mg(s) + F_2(g) \rightarrow MgF_2(s).$$

In this process, Mg atoms lose two electrons each to become $Mg^{2+}$ ions in $MgF_2$: Mg is oxidized. Each F atom of $F_2$ gains one electron to become a $F^-$ ion, for a total of two electrons gained for each $F_2$ molecule: $F_2$ is reduced.

$$Mg \rightarrow Mg^{2+} + 2e^-$$

$$2(F + e^- \rightarrow F^-)$$

14. In general, a synthesis reaction represents the reaction of elements or simple compounds to produce more complex substances. There are many examples of synthesis reactions, for example

$$N_2(g) + 3H_2(g) \rightarrow 2NH_3(g)$$

$$NaOH(aq) + CO_2(g) \rightarrow NaHCO_3(s).$$

Review:   Chapters 6 and 7

Decomposition reactions represent the breakdown of a more complex substance into simpler substances. There are many examples of decomposition reactions, for example

$$2H_2O_2(aq) \rightarrow 2H_2O(l) + O_2(g).$$

Synthesis and decomposition reactions are very often also oxidation–reduction reactions, especially if an elemental substance reacts or is generated. It is not necessary, however, for synthesis and decomposition reactions to always involve oxidation–reduction. The reaction between NaOH and $CO_2$ given as an example of a synthesis reaction does *not* represent oxidation–reduction.

16.  a.  $C(s) + O_2(g) \rightarrow CO_2(g)$

 b.  $2C(s) + O_2(g) \rightarrow 2CO(g)$

 c.  $2Li(l) + 2C(s) \rightarrow Li_2C_2(s)$

 d.  $FeO(s) + C(s) \rightarrow Fe(l) + CO(g)$

 e.  $C(s) + 2F_2(g) \rightarrow CF_4(g)$

18.  a.  $Ba(NO_3)_2(aq) + K_2CrO_4(aq) \rightarrow BaCrO_4(s) + 2KNO_3(aq)$

 b.  $NaOH(aq) + CH_3COOH(aq) \rightarrow H_2O(l) + NaCH_3COO(aq)$ (then evaporate the water from the solution)

 c.  $AgNO_3(aq) + NaCl(aq) \rightarrow AgCl(s) + NaNO_3(aq)$

 d.  $Pb(NO_3)_2(aq) + H_2SO_4(aq) \rightarrow PbSO_4(s) + 2HNO_3(aq)$

 e.  $2NaOH(aq) + H_2SO_4(aq) \rightarrow Na_2SO_4(aq) + 2H_2O(l)$ (then evaporate the water from the solution)

 f.  $Ba(NO_3)_2(aq) + Na_2CO_3(aq) \rightarrow BaCO_3(s) + 2NaNO_3(aq)$

20.  a.  $FeO(s) + 2HNO_3(aq) \rightarrow Fe(NO_3)_2(aq) + H_2O(l)$

   acid–base, double-displacement

 b.  $2Mg(s) + 2CO_2(g) + O_2(g) \rightarrow 2MgCO_3(s)$

   synthesis; **oxidation–reduction**

 c.  $2NaOH(s) + CuSO_4(aq) \rightarrow Cu(OH)_2(s) + Na_2SO_4(aq)$

   precipitation, double-displacement

 d.  $HI(aq) + KOH(aq) \rightarrow KI(aq) + H_2O(l)$

   acid–base, double–displacement

 e.  $C_3H_8(g) + 5O_2(g) \rightarrow 3CO_2(g) + 4H_2O(g)$

   combustion; **oxidation–reduction**

 f.  $Co(NH_3)_6Cl_2(s) \rightarrow CoCl_2(s) + 6NH_3(g)$

   decomposition

 g.  $2HCl(aq) + Pb(C_2H_3O_2)_2(aq) \rightarrow 2HC_2H_3O_2(aq) + PbCl_2(s)$

   precipitation, double-displacement

h. $C_{12}H_{22}O_{11}(s) \rightarrow 12C(s) + 11H_2O(g)$

decomposition; **oxidation–reduction**

i. $2Al(s) + 6HNO_3(aq) \rightarrow 2Al(NO_3)_3(aq) + 3H_2(g)$

single-displacement; **oxidation–reduction**

j. $4B(s) + 3O_2(g) \rightarrow 2B_2O_3(s)$

synthesis; **oxidation–reduction**

22. Specific examples will depend on students' responses. The following are general equations that illustrate each type of reaction:

precipitation: typical when solutions of two ionic solutes are mixed, and one of the new combinations of ions is insoluble.

$A^+B^-(aq) + C^+D^-(aq) \rightarrow AD(s) + C^+D^-(aq)$

single displacement: one element replaces a less reactive element from a compound.

$A(s) + B^+C^-(aq) \rightarrow A^+C^-(aq) + B(s)$

combustion: a rapid oxidation reaction, most commonly involving $O_2(g)$. Most examples in the text involve the combustion of hydrocarbons or hydrocarbon derivatives.

(hydrocarbon or derivative) + $O_2(g) \rightarrow CO_2(g) + H_2O(g)$

synthesis: elements or simple compounds combine to make more complicated molecules.

$A(s) + B(s) \rightarrow AB(s)$

oxidation–reduction: reactions in which electrons are transferred from one species to another. Examples of oxidation–reduction reactions include single-displacement, combustion, synthesis, and decomposition reactions.

decomposition: a compound breaks down into elements and/or simpler compounds.

$AB(s) \rightarrow A(s) + B(s)$

acid–base neutralization: a neutralization takes place when a proton from an acid combines with a hydroxide ion from a base to make a water molecule.

24. a. no reaction (all combinations are soluble)

b. $Ca^{2+}(aq) + SO_4^{2-}(aq) \rightarrow CaSO_4(s)$

c. $Pb^{2+}(aq) + S^{2-}(aq) \rightarrow PbS(s)$

d. $2Fe^{3+}(aq) + 3CO_3^{2-}(aq) \rightarrow Fe_2(CO_3)_3(s)$

e. $Hg_2^{2+}(aq) + 2Cl^-(aq) \rightarrow Hg_2Cl_2(s)$

f. $Ag^+(aq) + Cl^-(aq) \rightarrow AgCl(s)$

g. $3Ca^{2+}(aq) + 2PO_4^{3-}(aq) \rightarrow Ca_3(PO_4)_2(s)$

Since phosphoric acid is not a very strong acid, a more realistic equation might be

$3Ca^{2+}(aq) + 2H_3PO_4(aq) \rightarrow Ca_3(PO_4)_2(s) + 6H^+(aq)$

h. no reaction (all combinations are soluble)

# CHAPTER 8

# Chemical Composition

2. The empirical formula is CFH from the structure given. The empirical formula represents the smallest whole number ratio of the number and types of atoms present.

4. The average atomic mass takes into account the various isotopes of an element and the relative abundances in which those isotopes are found.

6. a. $40.08 \text{ amu Ca} \times \dfrac{1 \text{ Ca atom}}{40.08 \text{ amu}} = 1 \text{ Ca atom}$

   b. $919.5 \text{ amu W} \times \dfrac{1 \text{ W atom}}{183.9 \text{ amu}} = 5 \text{ W atoms}$

   c. $549.4 \text{ amu Mn} \times \dfrac{1 \text{ Mn atom}}{54.94 \text{ amu}} = 10 \text{ Mn atoms}$

   d. $6345 \text{ amu I} \times \dfrac{1 \text{ I atom}}{126.9 \text{ amu}} = 50 \text{ I atoms}$

   e. $2072 \text{ amu} \times \dfrac{1 \text{ Pb atom}}{207.2 \text{ amu}} = 10 \text{ Pb atoms}$

8. One copper atom has a mass of 63.55 amu.

   A sample containing 75 copper atoms would weigh: $75 \text{ atoms} \times \dfrac{63.55 \text{ amu}}{1 \text{ atom}} = 4766 \text{ amu}$;

   6100.8 amu of copper would represent: $6100.8 \text{ amu} \times \dfrac{1 \text{ atom}}{63.55 \text{ amu}} = 96 \text{ atoms}$.

10. 65.38 g (1.00 mol)

12. molar masses: Ag, 107.9 g; Cu, 63.55 g

    $300.0 \text{ g Ag} \times \dfrac{1 \text{ mol Ag}}{107.9 \text{ g Ag}} = 2.780 \text{ mol Ag}$

    $2.780 \text{ mol Ag} \times \dfrac{6.022 \times 10^{23}}{1 \text{ mol Ag}} = 1.674 \times 10^{24} \text{ atoms}$

    $2.780 \text{ mol Cu} \times \dfrac{63.55 \text{ g Cu}}{1 \text{ mol Cu}} = 176.7 \text{ g Cu}$

Chapter 8: Chemical Composition

14. The ratio of the atomic mass of Co to the atomic mass of F is (58.93 amu Co/19.00 amu F), and the mass of cobalt is given by

$$57.0 \text{ g F} \times \frac{58.93 \text{ amu Co}}{19.00 \text{ amu F}} = 177 \text{ g Co.}$$

16. mass of a chlorine atom = $3.16 \times 10^{-23}$ g $\times \dfrac{35.45 \text{ amu Cl}}{19.00 \text{ amu F}} = 5.90 \times 10^{-23}$ g

18. 0.25 mol Xe atoms $\times \dfrac{131.3 \text{ g Xe}}{1 \text{ mol Xe}} = 32.83$ g Xe = 33 g Xe (two significant figures)

   2.0 mol C atoms $\times \dfrac{12.01 \text{ g C}}{1 \text{ mol C}} = 24.02$ g C = 24 g C (two significant figures)

   The carbon sample weighs less.

20. a. $49.2 \text{ g S} \times \dfrac{1 \text{ mol}}{32.07 \text{ g}} = 1.53$ mol of S

   b. $7.44 \times 10^4 \text{ kg Pb} \times \dfrac{1000 \text{ g}}{1 \text{ kg}} \times \dfrac{1 \text{ mol}}{207.2 \text{ g}} = 3.59 \times 10^5$ mol Pb

   c. $3.27 \text{ mg Cl} \times \dfrac{1 \text{ g}}{1000 \text{ mg}} \times \dfrac{1 \text{ mol}}{35.45 \text{ g}} = 9.22 \times 10^{-5}$ mol Cl

   d. $4.01 \text{ g Li} \times \dfrac{1 \text{ mol}}{6.9419 \text{ g}} = 0.578$ mol Li

   e. $100.0 \text{ g Cu} \times \dfrac{1 \text{ mol}}{63.55 \text{ g}} = 1.574$ mol Cu

   f. $82.6 \text{ mg Sr} \times \dfrac{1 \text{ g}}{1000 \text{ mg}} \times \dfrac{1 \text{ mol}}{87.62 \text{ g}} = 9.43 \times 10^{-4}$ mol Sr

22. a. $0.00552 \text{ mol Ca} \times \dfrac{40.08 \text{ g}}{1 \text{ mol}} = 0.221$ g Ca

   b. $6.25 \text{ millimol B} \times \dfrac{1 \text{ mol}}{10^3 \text{ millmol}} \times \dfrac{10.81 \text{ g}}{1 \text{ mol}} = 0.0676$ g B

   c. $135 \text{ mol Al} \times \dfrac{26.98 \text{ g}}{1 \text{ mol}} = 3.64 \times 10^3$ g Al

   d. $1.34 \times 10^{-7} \text{ mol Ba} \times \dfrac{137.3 \text{ g}}{1 \text{ mol}} = 1.84 \times 10^{-5}$ g Ba

   e. $2.79 \text{ mol P} \times \dfrac{30.97 \text{ g}}{1 \text{ mol}} = 86.4$ g P

Chapter 8:   Chemical Composition

f.   $0.0000997 \text{ mol As} \times \dfrac{74.92 \text{ g}}{1 \text{ mol}} = 7.47 \times 10^{-3} \text{ g As}$

24.  a.   $125 \text{ Fe atoms} \times \dfrac{55.85 \text{ g Fe}}{6.022 \times 10^{23} \text{ Fe atoms}} = 1.16 \times 10^{-20} \text{ g}$

b.   $125 \text{ Fe atoms} \times \dfrac{55.85 \text{ amu}}{1 \text{ Fe atom}} = 6.98 \times 10^{3} \text{ amu}$

c.   $125 \text{ g Fe} \times \dfrac{1 \text{ mol Fe}}{55.85 \text{ g Fe}} = 2.24 \text{ mol Fe}$

d.   $125 \text{ mol Fe} \times \dfrac{55.85 \text{ g Fe}}{1 \text{ mol Fe}} = 6.98 \times 10^{3} \text{ g Fe}$

e.   $125 \text{ g Fe} \times \dfrac{6.022 \times 10^{23} \text{ Fe atoms}}{55.85 \text{ g Fe}} = 1.35 \times 10^{24} \text{ Fe atoms}$

f.   $125 \text{ mol Fe} \times \dfrac{6.022 \times 10^{23} \text{ Fe atoms}}{1 \text{ mol Fe}} = 7.53 \times 10^{25} \text{ Fe atoms}$

26.  (Answers will vary.) The molar mass is calculated by summing the individual atomic masses of the atoms in the formula. In the compound $CH_4$, the atomic mass of carbon and the atomic mass of four hydrogens are summed (giving a molar mass of 16.042 g/mol).

28.  a.   $KHCO_3$ potassium hydrogen carbonate, potassium bicarbonate

mass of 1 mol K = 39.10 g

mass of 1 mol H = 1.008 g

mass of 1 mol C = 12.01 g

mass of 3 mol O = 3(16.00 g) = 48.00 g

molar mass of $KHCO_3$ = (39.10 + 1.008 + 12.01 + 48.00) = 100.12 g

b.   $Hg_2Cl_2$ mercurous chloride, mercury(I) chloride

mass of 2 mol Hg = 2(200.6 g) = 401.2 g

mass of 2 mol Cl = 2(35.45 g) = 70.90 g

molar mass of $Hg_2Cl_2$ = (401.2 g + 70.90 g) = 472.1 g

c.   $H_2O_2$ hydrogen peroxide

mass of 2 mol H = 2(1.008 g) = 2.016 g

mass of 2 mol O = 2(16.00 g) = 32.00 g

molar mass of $H_2O_2$ = (2.016 g + 32.00 g) = 34.02 g

d.   $BeCl_2$ beryllium chloride

mass of 1 mol Be = 9.012 g

mass of 2 mol Cl = 2(35.45 g) = 70.90 g

molar mass of $BeCl_2$ = (9.012 g + 70.90 g) = 79.91 g

Chapter 8: Chemical Composition

  e. $Al_2(SO_4)_3$ aluminum sulfate

    mass of 2 mol Al = 2(26.98 g) = 53.96 g

    mass of 3 mol S = 3(32.07 g) = 96.21 g

    mass of 12 mol O = 12(16.00 g) = 192.0 g

    molar mass of $Al_2(SO_4)_3$ = (53.96 g + 96.21 g + 192.0 g) = 342.2 g

  f. $KClO_3$ potassium chlorate

    mass of 1 mol K = 39.10 g

    mass of 1 mol Cl = 35.45 g

    mass of 3 mol O = 3(16.00 g) = 48.00 g

    molar mass of $KClO_3$ = 122.55 g

30. a. $CO_2$

    mass of 1 mol C = 12.01 g

    mass of 2 mol O = 2(16.00 g) = 32.00 g

    molar mass of $CO_2$ = (12.01 g + 32.00 g) = 44.01 g

  b. $AlPO_4$

    mass of 1 mol Al = 26.98 g

    mass of 1 mol P = 30.97 g

    mass of 4 mol O = 4(16.00 g) = 64.00 g

    molar mass of $AlPO_4$ = (26.98 g + 30.97 g + 64.00 g) = 121.95 g

  c. $Fe_2(CO_3)_3$

    mass of 2 mol Fe = 2(55.85 g) = 111.70 g

    mass of 3 mol C = 3(12.01 g) = 36.03 g

    mass of 9 mol O = 9(16.00 g) = 144.00 g

    molar mass of $Fe_2(CO_3)_3$ = (111.70 g + 36.03 g + 144.00 g) = 291.73 g

  d. $Pb(NO_3)_2$

    mass of 1 mol Pb = 207.2 g

    mass of 2 mol N = 2(14.01 g) = 28.02 g

    mass of 6 mol O = 6(16.00 g) = 96.00 g

    molar mass of $Pb(NO_3)_2$ = (207.2 g + 28.02 g + 96.00 g) = 331.2 g

  e. $SrCl_2$

    mass of 1 mol Sr = 87.62 g

    mass of 2 mol Cl = 2(35.45 g) = 70.90 g

    molar mass of $SrCl_2$ = (87.62 g + 70.90 g) = 158.52 g

## Chapter 8: Chemical Composition

32. a. molar mass $Al_2O_3$ = 101.96 g

   $47.2 \text{ g} \times \dfrac{1 \text{ mol}}{101.96 \text{ g}} = 0.463 \text{ mol}$

   b. molar mass KBr = 119.00 g

   $1.34 \text{ kg} \times \dfrac{1000 \text{ g}}{1 \text{ kg}} \times \dfrac{1 \text{ mol}}{119.00 \text{ g}} = 11.3 \text{ mol}$

   c. molar mass Ge = 72.59 g

   $521 \text{ mg} \times \dfrac{1 \text{ g}}{1000 \text{ mg}} \times \dfrac{1 \text{ mol}}{72.59 \text{ g}} = 7.18 \times 10^{-3} \text{ mol}$

   d. molar mass of U = 238.0 g

   $56.2 \text{ μg} \times \dfrac{1 \text{ g}}{10^6 \text{ μg}} \times \dfrac{1 \text{ mol}}{238.0 \text{ g}} = 2.36 \times 10^{-7} \text{ mol}$

   e. molar mass of $NaC_2H_3O_2$ = 82.03 g

   $29.7 \text{ g} \times \dfrac{1 \text{ mol}}{82.03 \text{ g}} = 1.69 \text{ mol} = 0.362 \text{ mol}$

   f. molar mass of $SO_3$ = 80.07 g

   $1.03 \text{ g} \times \dfrac{1 \text{ mol}}{80.07 \text{ g}} = 0.0129 \text{ mol}$

34. a. molar mass of $Li_2CO_3$ = 73.89 g

   $1.95 \times 10^{-3} \text{ g} \times \dfrac{1 \text{ mol}}{73.89 \text{ g}} = 2.64 \times 10^{-5} \text{ mol}$

   b. molar mass of $CaCl_2$ = 110.98 g

   $4.23 \text{ kg} \times \dfrac{1000 \text{ g}}{1 \text{ kg}} \times \dfrac{1 \text{ mol}}{110.98 \text{ g}} = 38.1 \text{ mol}$

   c. molar mass of $SrCl_2$ = 158.52 g

   $1.23 \text{ mg} \times \dfrac{1 \text{ g}}{1000 \text{ mg}} \times \dfrac{1 \text{ mol}}{158.52 \text{ g}} = 7.76 \times 10^{-6} \text{ mol}$

   d. molar mass of $CaSO_4$ = 136.15 g

   $4.75 \text{ g} \times \dfrac{1 \text{ mol}}{136.15 \text{ g}} = 3.49 \times 10^{-2} \text{ mol}$

   e. molar mass of $NO_2$ = 46.01 g

   $96.2 \text{ mg} \times \dfrac{1 \text{ g}}{1000 \text{ mg}} \times \dfrac{1 \text{ mol}}{46.01 \text{ g}} = 2.09 \times 10^{-3} \text{ mol}$

Chapter 8: Chemical Composition

f. molar mass of $Hg_2Cl_2$ = 472.1 g

$$12.7 \text{ g} \times \frac{1 \text{ mol}}{472.1 \text{ g}} = 0.0269 \text{ mol}$$

36. a. molar mass of $SO_3$ = 80.07 g

$$6.14 \times 10^{-4} \text{ mol} \times \frac{80.07 \text{ g}}{1 \text{ mol}} = 0.0492 \text{ g}$$

b. molar mass of $PbO_2$ = 239.2 g

$$3.11 \times 10^5 \text{ mol} \times \frac{239.2 \text{ g}}{1 \text{ mol}} = 7.44 \times 10^7 \text{ g}$$

c. molar mass of $CHCl_3$ = 119.368 g

$$0.495 \text{ mol} \times \frac{119.368 \text{ g}}{1 \text{ mol}} = 59.1 \text{ g}$$

d. molar mass of $C_2H_3Cl_3$ = 133.394 g

$$2.45 \times 10^{-8} \text{ mol} \times \frac{133.394 \text{ g}}{1 \text{ mol}} = 3.27 \times 10^{-6} \text{ g}$$

e. molar mass of LiOH = 23.949 g

$$0.167 \text{ mol} \times \frac{23.949 \text{ g}}{1 \text{ mol}} = 4.00 \text{ g}$$

f. molar mass of CuCl = 99.00 g

$$5.26 \text{ mol} \times \frac{99.00 \text{ g}}{1 \text{ mol}} = 521 \text{ g}$$

38. a. molar mass $C_6H_6$ = 78.11 g

$$0.994 \text{ mol} \times \frac{78.11 \text{ g}}{1 \text{ mol}} = 77.6 \text{ g}$$

b. molar mass $CaH_2$ = 42.10 g

$$4.21 \text{ mol} \times \frac{42.10 \text{ g}}{1 \text{ mol}} = 177 \text{ g}$$

c. molar mass $H_2O_2$ = 34.02 g

$$1.79 \times 10^{-4} \text{ mol} \times \frac{34.02 \text{ g}}{1 \text{ mol}} = 6.09 \times 10^{-3} \text{ g}$$

d. molar mass $C_6H_{12}O_6$ = 180.16 g

$$1.22 \text{ mmol} \times \frac{1 \text{ mol}}{10^3 \text{ mmol}} \times \frac{105.99 \text{ g}}{1 \text{ mol}} = 0.220 \text{ g}$$

Chapter 8: Chemical Composition

    e.    molar mass Sn = 118.7 g

$$10.6 \text{ mol} \times \frac{118.7 \text{ g}}{1 \text{ mol}} = 1.26 \times 10^3 \text{ g}$$

    f.    molar mass $SrF_2$ = 125.62 g

$$0.000301 \text{ mol} \times \frac{125.62 \text{ g}}{1 \text{ mol}} = 0.0378 \text{ g}$$

40.    a.    $3.54 \text{ mol } SO_2 \times \dfrac{6.022 \times 10^{23} \text{ molecules}}{1 \text{ mol}} = 2.13 \times 10^{24} \text{ molecules } SO_2$

    b.    molar mass of $SO_2$ = 64.07 g

$$3.54 \text{ g} \times \frac{6.022 \times 10^{23} \text{ molecules}}{64.07 \text{ g}} = 3.33 \times 10^{22} \text{ molecules } SO_2$$

    c.    molar mass of $NH_3$ = 17.034 g

$$4.46 \times 10^{-5} \text{ g} \times \frac{6.022 \times 10^{23} \text{ molecules}}{17.034 \text{ g}} = 1.58 \times 10^{18} \text{ molecules } NH_3$$

    d.    $4.46 \times 10^{-5} \text{ mol } NH_3 \times \dfrac{6.022 \times 10^{23} \text{ molecules}}{1 \text{ mol}} = 2.69 \times 10^{19} \text{ molecules } NH_3$

    e.    molar mass of $C_2H_6$ = 30.068 g

$$1.96 \text{ mg} \times \frac{1 \text{ g}}{1000 \text{ mg}} \times \frac{1 \text{ mol}}{30.068 \text{ g}} \times \frac{6.022 \times 10^{23} \text{ molecules}}{1 \text{ mol}} = 3.93 \times 10^{19} \text{ molecules } C_2H_6$$

42.    a.    molar mass of $Na_2SO_4$ = 142.1 g

$$2.01 \text{ g } Na_2SO_4 \times \frac{1 \text{ mol } Na_2SO_4}{142.1 \text{ g}} \times \frac{1 \text{ mol S}}{1 \text{ mol } Na_2SO_4} = 0.0141 \text{ mol S}$$

    b.    molar mass of $Na_2SO_3$ = 126.1 g

$$2.01 \text{ g } Na_2SO_3 \times \frac{1 \text{ mol } Na_2SO_3}{126.1 \text{ g}} \times \frac{1 \text{ mol S}}{1 \text{ mol } Na_2SO_3} = 0.0159 \text{ mol S}$$

    c.    molar mass of $Na_2S$ = 78.05 g

$$2.01 \text{ g } Na_2S \times \frac{1 \text{ mol } Na_2S}{78.05 \text{ g}} \times \frac{1 \text{ mol S}}{1 \text{ mol } Na_2S} = 0.0258 \text{ mol S}$$

    d.    molar mass of $Na_2S_2O_3$ = 158.1 g

$$2.01 \text{ g } Na_2S_2O_3 \times \frac{1 \text{ mol } Na_2S_2O_3}{158.1 \text{ g}} \times \frac{2 \text{ mol S}}{1 \text{ } Na_2S_2O_3} = 0.0254 \text{ mol S}$$

44.    The percent composition of each element in the compound does not change because a compound consists of the same percent composition by mass regardless of the starting amount in the sample.

46. a.  mass of Zn present = 65.38 g
        mass of O present = 16.00 g
        molar mass of ZnO = 81.38 g

        $\%Zn = \dfrac{65.38 \text{ g Zn}}{81.38 \text{ g}} \times 100 = 80.34\% \text{ Zn}$

        $\%O = \dfrac{16.00 \text{ g O}}{81.38 \text{ g}} \times 100 = 19.66\% \text{ O}$

   b.  mass of Na present = 2(22.99 g) = 45.98 g
       mass of S present = 32.07 g
       molar mass of $Na_2S$ = 78.05 g

       $\%Na = \dfrac{45.98 \text{ g Na}}{78.05 \text{ g}} \times 100 = 58.91\% \text{ Na}$

       $\%S = \dfrac{32.07 \text{ g S}}{78.05 \text{ g}} \times 100 = 41.09\% \text{ S}$

   c.  mass of Mg present = 24.31 g
       mass of O present = 2(16.00 g) = 32.00 g
       mass of H present = 2(1.008 g) = 2.016 g
       molar mass of $Mg(OH)_2$ = 58.33 g

       $\%Mg = \dfrac{24.31 \text{ g Mg}}{58.33 \text{ g}} \times 100 = 41.68\% \text{ Mg}$

       $\%O = \dfrac{32.00 \text{ g O}}{58.33 \text{ g}} \times 100 = 54.86\% \text{ O}$

       $\%H = \dfrac{2.016 \text{ g H}}{58.33 \text{ g}} \times 100 = 3.456\% \text{ H}$

   d.  mass of H present = 2(1.008 g) = 2.016 g
       mass of O present = 2(16.00 g) = 32.00 g
       molar mass of $H_2O_2$ = 34.02 g

       $\%H = \dfrac{2.016 \text{ g H}}{34.02 \text{ g}} \times 100 = 5.926\% \text{ H}$

       $\%O = \dfrac{32.00 \text{ g O}}{34.02 \text{ g}} \times 100 = 94.06\% \text{ O}$

   e.  mass of Ca present = 40.08 g
       mass of H present = 2(1.008 g) = 2.016 g
       molar mass of $CaH_2$ = 42.10 g

Chapter 8: Chemical Composition

$$\%Ca = \frac{40.08 \text{ g Ca}}{42.10 \text{ g}} \times 100 = 95.20\% \text{ Ca}$$

$$\%H = \frac{2.016 \text{ g H}}{42.10 \text{ g}} \times 100 = 4.789\% \text{ H}$$

f. mass of K present = 2(39.10 g) = 78.20 g

mass of O present = 16.00 g

molar mass of $K_2O$ = 94.20 g

$$\%K = \frac{78.20 \text{ g K}}{94.20 \text{ g}} \times 100 = 83.01\% \text{ K}$$

$$\%O = \frac{16.00 \text{ g O}}{94.20 \text{ g}} \times 100 = 16.99\% \text{ O}$$

48. a. molar mass of $CuBr_2$ = 223.35 g

$$\% \text{ Cu} = \frac{63.55 \text{ g Cu}}{223.35 \text{ g}} \times 100 = 28.45\% \text{ Cu}$$

b. molar mass of CuBr = 143.45 g

$$\% \text{ Cu} = \frac{63.55 \text{ g Cu}}{143.45 \text{ g}} \times 100 = 44.30\% \text{ Cu}$$

c. molar mass of $FeCl_2$ = 126.75 g

$$\% \text{ Fe} = \frac{55.85 \text{ g Fe}}{126.75 \text{ g}} \times 100 = 44.06\% \text{ Fe}$$

d. molar mass of $FeCl_3$ = 162.2 g

$$\% \text{ Fe} = \frac{55.85 \text{ g Fe}}{162.2 \text{ g}} \times 100 = 34.43\% \text{ Fe}$$

e. molar mass of $CoI_2$ = 312.73 g

$$\% \text{ Co} = \frac{58.93 \text{ g Co}}{312.73 \text{ g}} \times 100 = 18.84\% \text{ Co}$$

f. molar mass of $CoI_3$ = 439.63 g

$$\% \text{ Co} = \frac{58.93 \text{ g Co}}{439.63 \text{ g}} \times 100 = 13.40\% \text{ Co}$$

g. molar mass of SnO = 134.7 g

$$\% \text{ Sn} = \frac{118.7 \text{ g Sn}}{134.7 \text{ g}} \times 100 = 88.12\% \text{ Sn}$$

Chapter 8: Chemical Composition

    h.    molar mass of $SnO_2$ = 150.7 g

$$\% \text{ Sn} = \frac{118.7 \text{ g Sn}}{150.7 \text{ g}} \times 100 = 78.77\% \text{ Sn}$$

50.    a.    molar mass of $CO_2$ = 44.01 g

$$\% \text{ O} = \frac{32.00 \text{ g O}}{44.01 \text{ g}} \times 100\% = 72.71\% \text{ O}$$

    b.    molar mass of $NaNO_3$ = 85.00 g

$$\% \text{ O} = \frac{48.00 \text{ g O}}{85.00 \text{ g}} \times 100\% = 56.47\% \text{ O}$$

    c.    molar mass of $FePO_4$ = 150.82 g

$$\% \text{ O} = \frac{64.00 \text{ g O}}{150.82 \text{ g}} \times 100\% = 42.43\% \text{ O}$$

    d.    molar mass of $(NH_4)_2CO_3$ = 96.094 g

$$\% \text{ O} = \frac{48.00 \text{ g O}}{96.094 \text{ g}} \times 100\% = 49.95\% \text{ O}$$

    e.    molar mass of $Al_2(SO_4)_3$ = 342.14 g

$$\% \text{ O} = \frac{192.00 \text{ g}}{342.14 \text{ g}} \times 100\% = 56.117\% \text{ O}$$

52.    a.    molar mass of $(NH_4)_2S$ = 68.15 g; molar mass of $S^{2-}$ ion = 32.07 g

$$\% \text{ S}^{2-} = \frac{32.07 \text{ g S}^{2-}}{68.15 \text{ g NH}_4\text{S}} \times 100 = 47.06\% \text{ S}^{2-}$$

    b.    molar mass of $CaCl_2$ = 110.98 g; molar mass of $Cl^-$ = 35.45 g

$$\% \text{ Cl}^- = \frac{70.90 \text{ g Cl}^-}{110.98 \text{ g CaCl}_2} \times 100 = 63.89\% \text{ Cl}^-$$

    c.    molar mass of BaO = 153.3 g; molar mass of $O^{2-}$ ion = 16.00 g

$$\% \text{ O}^{2-} = \frac{16.00 \text{ g O}^{2-}}{153.3 \text{ g BaO}} \times 100 = 10.44\% \text{ O}^{2-}$$

    d.    molar mass of $NiSO_4$ = 154.76 g; molar mass of $SO_4^{2-}$ ion = 96.07 g

$$\% \text{ SO}_4^{2-} = \frac{96.07 \text{ g SO}_4^{2-}}{154.76 \text{ g NiSO}_4} \times 100 = 62.08\% \text{ SO}_4^{2-}$$

54.    The empirical formula indicates the smallest whole number ratio of the number and type of atoms present in a molecule. For example, $NO_2$ and $N_2O_4$ both have two oxygen atoms for every nitrogen atom and therefore have the same empirical formula

Chapter 8:     Chemical Composition

56. a. yes (each of these has empirical formula CH)
    b. no (the number of hydrogen atoms is wrong)
    c. yes (both have empirical formula $NO_2$)
    d. no (the number of hydrogen and oxygen atoms is wrong)

58. Assume we have 100.0 g of the compound so that the percentages become masses.

    $11.64 \text{ g N} \times \dfrac{1 \text{ mol}}{14.01 \text{ g}} = 0.8308 \text{ mol N}$

    $88.36 \text{ g Cl} \times \dfrac{1 \text{ mol}}{35.45 \text{ g}} = 2.493 \text{ mol Cl}$

    Dividing both of these numbers of moles by the smaller number of moles gives

    $\dfrac{0.8308 \text{ mol N}}{0.8308} = 1.000 \text{ mol N}$

    $\dfrac{2.493 \text{ mol Cl}}{0.8308 \text{ mol}} = 3.001 \text{ mol Cl}$

    The empirical formula is $NCl_3$.

60. Assume we have 100.0 g of the compound, so that the percentages become masses.

    $78.14 \text{ g B} \times \dfrac{1 \text{ mol}}{10.81 \text{ g}} = 7.228 \text{ mol B}$

    $21.86 \text{ g H} \times \dfrac{1 \text{ mol}}{1.008 \text{ g}} = 21.69 \text{ mol H}$

    Dividing each number of moles by the smaller number of moles gives

    $\dfrac{7.228 \text{ mol B}}{7.228 \text{ mol}} = 1.000 \text{ mol B}$

    $\dfrac{21.69 \text{ mol H}}{7.228 \text{ mol}} = 3.000 \text{ mol H}$

    The empirical formula is $BH_3$.

62. Consider 100.0 g of the compound so that percentages become masses.

    $45.56 \text{ g Sn} \times \dfrac{1 \text{ mol}}{118.7 \text{ g}} = 0.3838 \text{ mol Sn}$

    $54.43 \text{ g Cl} \times \dfrac{1 \text{ mol}}{35.45 \text{ g}} = 1.535 \text{ mol Cl}$

    Dividing each number of moles by the smaller number of moles gives

    $\dfrac{0.3838 \text{ mol Sn}}{0.3838 \text{ mol}} = 1.000 \text{ mol Sn}$

$$\frac{1.535 \text{ mol Cl}}{0.3838 \text{ mol}} = 3.999 \text{ mol Cl}$$

The empirical formula is SnCl₄.

64. Consider 100.0 g of the compound.

$$55.06 \text{ g Co} \times \frac{1 \text{ mol}}{58.93 \text{ g}} = 0.9343 \text{ mol Co}$$

If the sulfide of cobalt is 55.06% Co, then it is 44.94% S by mass.

$$44.94 \text{ g S} \times \frac{1 \text{ mol}}{32.07 \text{ g}} = 1.401 \text{ mol S}$$

Dividing each number of moles by the smaller (0.9343 mol Co) gives

$$\frac{0.09343 \text{ mol Co}}{0.9343} = 1.000 \text{ mol Co}$$

$$\frac{1.401 \text{ mol S}}{0.9343 \text{ mol}} = 1.500 \text{ mol S}$$

Multiplying by two, to convert to whole numbers of moles, gives the empirical formula for the compound as Co₂S₃.

66. $2.50 \text{ g Al} \times \frac{1 \text{ mol}}{26.98 \text{ g}} = 0.09266 \text{ mol Al}$

$$5.28 \text{ g F} \times \frac{1 \text{ mol}}{19.00 \text{ g}} = 0.2779 \text{ mol F}$$

Dividing each number of moles by the smaller number of moles gives

$$\frac{0.09266 \text{ mol Al}}{0.09266 \text{ mol}} = 1.000 \text{ mol Al}$$

$$\frac{0.2779 \text{ mol F}}{0.09266 \text{ mol}} = 2.999 \text{ mol F}$$

The empirical formula is just AlF₃. Note the similarity between this problem and question 65: they differ in the way the data is given. In question 65, you were given the mass of the product, and first had to calculate how much fluorine had reacted.

68. Consider 100.0 g of the compound so that percentages become masses.

$$46.46 \text{ g Li} \times \frac{1 \text{ mol}}{6.941 \text{ g}} = 6.694 \text{ mol Li}$$

$$53.54 \text{ g O} \times \frac{1 \text{ mol}}{16.00 \text{ g}} = 3.346 \text{ mol O}$$

Dividing each number of moles by the smaller number of moles gives

Chapter 8: Chemical Composition

$$\frac{6.694 \text{ mol Li}}{3.346 \text{ mol}} = 2.001 \text{ mol Li}$$

$$\frac{3.346 \text{ mol O}}{3.346 \text{ mol}} = 1.000 \text{ mol O}$$

The empirical formula is $Li_2O$

70. Consider 100.0 g of the compound.

$$59.78 \text{ g Li} \times \frac{1 \text{ mol}}{6.941 \text{ g}} = 8.613 \text{ mol Li}$$

$$40.22 \text{ g N} \times \frac{1 \text{ mol}}{14.01 \text{ g}} = 2.871 \text{ mol N}$$

Dividing each number of moles by the smaller number of moles (2.871 mol N) gives

$$\frac{8.613 \text{ mol Li}}{2.871 \text{ mol}} = 3.000 \text{ mol Li}$$

$$\frac{2.871 \text{ mol N}}{2.871 \text{ mol}} = 1.000 \text{ mol N}$$

The empirical formula is $Li_3N$.

72. Consider 100.00 g of the compound so that percentages become masses.

$$85.96 \text{ g C} \times \frac{1 \text{ mol C}}{12.01 \text{ g}} = 7.157 \text{ mol C}$$

$$4.92 \text{ g H} \times \frac{1 \text{ mol H}}{1.008 \text{ g}} = 4.88 \text{ mol H}$$

$$(100.00 \text{ g} - 85.96 \text{ g} - 4.92 \text{ g} = 9.12 \text{ g N}) \times \frac{1 \text{ mol N}}{14.01 \text{ g}} = 0.651 \text{ mol N}$$

Dividing each number of moles by the smaller number of moles gives

$$\frac{7.157 \text{ mol C}}{0.651 \text{ mol}} = 11.0 \text{ mol C}$$

$$\frac{4.88 \text{ mol H}}{0.651 \text{ mol}} = 7.50 \text{ mol H}$$

$$\frac{0.651 \text{ mol N}}{0.651 \text{ mol}} = 1.00 \text{ mol N}$$

Multiplying these relative numbers of moles by 2 to give whole numbers gives the empirical formula as $C_{22}H_{15}N_2$.

Chapter 8: Chemical Composition

74. Compound 1: Assume 100.0 g of the compound.

$$22.55 \text{ g P} \times \frac{1 \text{ mol}}{30.97 \text{ g}} = 0.7281 \text{ mol P}$$

$$77.45 \text{ g Cl} \times \frac{1 \text{ mol}}{35.45 \text{ g}} = 2.185 \text{ mol Cl}$$

Dividing each number of moles by the smaller (0.7281 mol P) indicates that the formula of Compound 1 is $PCl_3$.

Compound 2: Assume 100.0 g of the compound.

$$14.87 \text{ g P} \times \frac{1 \text{ mol}}{30.97 \text{ g}} = 0.4801 \text{ mol P}$$

$$85.13 \text{ g Cl} \times \frac{1 \text{ mol}}{35.45 \text{ g}} = 2.401 \text{ mol Cl}$$

Dividing each number of moles by the smaller (0.4801 mol P) indicates that the formula of Compound 2 is $PCl_5$.

76. molecular formula: $C_6H_{12}O_6$ (count the number of each type of element in the molecule represented); empirical formula: $CH_2O$ (simplest whole-number ratio of $C_6H_{12}O_6$; divisible by 6)

78. empirical formula mass of CH = 13 g

$$n = \frac{\text{molar mass}}{\text{empirical formula mass}} = \frac{78 \text{ g}}{13 \text{ g}} = 6$$

The molecular formula is $(CH)_6$ or $C_6H_6$.

80. empirical formula mass of $C_2H_5O$ = 46 g

$$n = \frac{\text{molar mass}}{\text{empirical formula mass}} = \frac{90 \text{ g}}{46 \text{ g}} = \sim 2$$

molecular formula is $(C_2H_5O)_2 = C_4H_{10}O_2$

82.

$$69.6 \text{ g S} \times \frac{1 \text{ mol S}}{32.07 \text{ g S}} = \frac{2.17 \text{ mol S}}{2.17 \text{ mol}} = 1$$

$$(100.0 - 69.6) \text{ g N} \times \frac{1 \text{ mol N}}{14.01 \text{ g N}} = \frac{2.17 \text{ mol N}}{2.17 \text{ mol}} = 1$$

The empirical formula is therefore SN, with a molar mass of 46.08 g/mol.

$$n = \frac{\text{molar mass}}{\text{empirical formula mass}} = \frac{184 \text{ g/mol}}{46.08 \text{ g/mol}} = 4$$

The molecular formula is $(SN)_4 = S_4N_4$. Therefore, the correct name for this compound is tetrasulfur tetranitride or tetranitrogen tetrasulfide (if NS is used as the empirical formula).

Chapter 8: Chemical Composition

84.  
| | | |
|---|---|---|
| 5.00 g Al | 0.185 mol | $1.12 \times 10^{23}$ atoms |
| 0.140 g Fe | 0.00250 mol | $1.51 \times 10^{21}$ atoms |
| $2.7 \times 10^2$ g Cu | 4.3 mol | $2.6 \times 10^{24}$ atoms |
| 0.00250 g Mg | $1.03 \times 10^{-4}$ mol | $6.19 \times 10^{19}$ atoms |
| 0.062 g Na | $2.7 \times 10^{-3}$ mol | $1.6 \times 10^{21}$ atoms |
| $3.95 \times 10^{-18}$ g U | $1.66 \times 10^{-20}$ mol | $1.00 \times 10^4$ atoms |

86. mass of 2 mol X = 2(41.2 g) = 82.4 g

mass of 1 mol Y = 57.7 g = 57.7 g

mass of 3 mol Z = 3(63.9 g) = 191.7 g

molar mass of $X_2YZ_3$ = 331.8 g

$\%X = \dfrac{82.4 \text{ g}}{331.8 \text{ g}} \times 100 = 24.8 \% X$

$\% Y = \dfrac{57.7 \text{ g}}{331.8 \text{ g}} \times 100 = 17.4\% \text{ Y}$

$\% Z = \dfrac{191.7 \text{ g}}{331.8 \text{ g}} \times 100 = 57.8\% \text{ Z}$

If the molecular formula were actually $X_4Y_2Z_6$, the percentage composition would be the same, and the *relative* mass of each element present would not change. The molecular formula is always a whole number multiple of the empirical formula.

88. For the first compound (*restricted* amount of oxygen)

$2.118 \text{ g Cu} \times \dfrac{1 \text{ mol}}{63.55 \text{ g}} = 0.03333 \text{ mol Cu}$

$0.2666 \text{ g O} \times \dfrac{1 \text{ mol}}{16.00 \text{ g}} = 0.01666 \text{ mol O}$

Since the number of moles of Cu (0.03333 mol) is twice the number of moles of O (0.01666 mol), the empirical formula is $Cu_2O$.

For the second compound (stream of pure oxygen)

$2.118 \text{ g Cu} \times \dfrac{1 \text{ mol}}{63.55 \text{ g}} = 0.03333 \text{ mol Cu}$

$0.5332 \text{ g O} \times \dfrac{1 \text{ mol}}{16.00 \text{ g}} = 0.03333 \text{ mol O}$

Since the numbers of moles are the same, the empirical formula is CuO.

90.  a.   molar mass $H_2O$ = 18.02 g

$4.21 \text{ g} \times \dfrac{1 \text{ mol}}{18.02 \text{ g}} \times \dfrac{6.022 \times 10^{23} \text{ molecules}}{1 \text{ mol}} = 1.41 \times 10^{23} \text{ molecules}$

The sample contains $1.41 \times 10^{23}$ oxygen atoms and $2(1.41 \times 10^{23}) = 2.82 \times 10^{23}$ hydrogen atoms.

b. molar mass $CO_2 = 44.01$ g

$$6.81 \text{ g} \times \frac{1 \text{ mol}}{44.01 \text{ g}} \times \frac{6.022 \times 10^{23} \text{ molecules}}{1 \text{ mol}} = 9.32 \times 10^{22} \text{ molecules}$$

The sample contains $9.32 \times 10^{22}$ carbon atoms and $2(9.32 \times 10^{22}) = 1.86 \times 10^{23}$ oxygen atoms.

c. molar mass $C_6H_6 = 78.11$ g

$$0.000221 \text{ g} \times \frac{1 \text{ mol}}{78.11 \text{ g}} \times \frac{6.022 \times 10^{23} \text{ molecules}}{1 \text{ mol}} = 1.70 \times 10^{18} \text{ molec.}$$

The sample contains $6(1.70 \times 10^{18}) = 1.02 \times 10^{19}$ atoms of each element.

d. $2.26 \text{ mol} \times \dfrac{6.022 \times 10^{23} \text{ molecules}}{1 \text{ mol}} = 1.36 \times 10^{24}$ molecules

atoms C = $12(1.36 \times 10^{24}) = 1.63 \times 10^{25}$ atoms

atoms H = $22(1.36 \times 10^{24}) = 2.99 \times 10^{25}$ atoms

atoms O = $11(1.36 \times 10^{24}) = 1.50 \times 10^{25}$ atoms

92. a. molar mass of $C_3O_2 = 3(12.01 \text{ g}) + 2(16.00 \text{ g}) = 68.03$ g

$$\% C = \frac{36.03 \text{ g C}}{68.03 \text{ g}} \times 100 = 52.96\% \text{ C}$$

$$7.819 \text{ g } C_3O_2 \times \frac{52.96 \text{ g C}}{100.0 \text{ g } C_3O_2} = 4.141 \text{ g C}$$

$$4.141 \text{ g C} \times \frac{6.022 \times 10^{23} \text{ molecules}}{12.01 \text{ g C}} = 2.076 \times 10^{23} \text{ C atoms}$$

b. molar mass of CO = $12.01 \text{ g} + 16.00 \text{ g} = 28.01$ g

$$\% C = \frac{12.01 \text{ g C}}{28.01 \text{ g}} \times 100 = 42.88\% \text{ C}$$

$$1.53 \times 10^{21} \text{ molecules CO} \times \frac{1 \text{ C atom}}{1 \text{ molecule CO}} = 1.53 \times 10^{21} \text{ C atoms}$$

$$1.53 \times 10^{21} \text{ C atoms} \times \frac{12.01 \text{ g C}}{6.022 \times 10^{23} \text{ C atoms}} = 0.0305 \text{ g C}$$

c. molar mass of $C_6H_6O = 6(12.01 \text{ g}) + 6(1.008 \text{ g}) + 16.00 \text{ g} = 94.11$ g

$$\% C = \frac{72.06 \text{ g C}}{94.11 \text{ g}} \times 100 = 76.57\% \text{ C}$$

Chapter 8:    Chemical Composition

$$0.200 \text{ mol } C_6H_6O \times \frac{6 \text{ mol C}}{1 \text{ mol } C_6H_6O} = 1.20 \text{ mol C}$$

$$1.20 \text{ mol C} \times \frac{12.01 \text{ g}}{1 \text{ mol}} = 14.4 \text{ g C}$$

$$14.4 \text{ g C} \times \frac{6.022 \times 10^{23} \text{ C atoms}}{12.01 \text{ g C}} = 7.22 \times 10^{23} \text{ C atoms}$$

94. (a); Since there are an equal number of atoms in each sample, there are also an equal number of moles (due to Avogadro's number). The molar mass of copper (63.55 g/mol) is more than twice as great as the molar mass of aluminum (26.98 g/mol).

96. a.

$$79.89 \text{ g C} \times \frac{1 \text{ mol C}}{12.01 \text{ g C}} = \frac{6.652 \text{ mol C}}{6.652 \text{ mol}} = 1$$

$$20.11 \text{ g H} \times \frac{1 \text{ mol H}}{1.008 \text{ g H}} = \frac{19.95 \text{ mol H}}{6.652 \text{ mol}} = 3$$

The empirical formula is therefore $CH_3$, with a molar mass of 15.034 g/mol (not worrying about significant figures).

b. (iv); 30.068 g/mol is the only molar mass listed that is a multiple of the empirical mass.

98. 153.8 g $CCl_4$ = 6.022 × 10²³ molecules $CCl_4$

$$1 \text{ molecule} \times \frac{153.8 \text{ g CCl}_4}{6.022 \times 10^{23} \text{ molecules}} = 2.554 \times 10^{-22} \text{ g}$$

100. (a); Both $NO_2$ and $F_2$ are molecules and since you have equal moles of each (but not necessarily 1 mole of each), the number of molecules must be the same (6.022 × 10²³ = 1 mol). However, the number of atoms is different since $NO_2$ contains 3 atoms and $F_2$ contains 2. Furthermore, the masses are different since $NO_2$ and $F_2$ have different molar masses.

102. (d); Since the charge on the transition metal ion is +3 and the charge on the oxygen ion is –2, the formula of the compound can be represented as $M_2O_3$. Thus, choose the transition metal (Ru) that gives a molar mass of 250.2 g/mol ($Ru_2O_3$).

104. We use the *average* mass because this average is a *weighted average* and takes into account both the masses and the relative abundances of the various isotopes.

106. $1.98 \times 10^{13}$ amu × $\dfrac{1 \text{ Na atom}}{22.99 \text{ amu}}$ = $8.61 \times 10^{11}$ Na atoms

$3.01 \times 10^{23}$ Na atoms × $\dfrac{22.99 \text{ amu}}{1 \text{ Na atom}}$ = $6.92 \times 10^{24}$ amu

108. a. 5.0 mol K × $\dfrac{39.10 \text{ g}}{1 \text{ mol}}$ = 195 g = $2.0 \times 10^2$ g K

Chapter 8: Chemical Composition

b. $0.000305 \text{ mol Hg} \times \dfrac{200.6 \text{ g}}{1 \text{ mol}} = 0.0612 \text{ g Hg}$

c. $2.31 \times 10^{-5} \text{ mol Mn} \times \dfrac{54.94 \text{ g}}{1 \text{ mol}} = 1.27 \times 10^{-3} \text{ g Mn}$

d. $10.5 \text{ mol P} \times \dfrac{30.97 \text{ g}}{1 \text{ mol}} = 325 \text{ g P}$

e. $4.9 \times 10^4 \text{ mol Fe} \times \dfrac{55.85 \text{ g}}{1 \text{ mol}} = 2.7 \times 10^6 \text{ g Fe}$

f. $125 \text{ mol Li} \times \dfrac{6.941 \text{ g}}{1 \text{ mol}} = 868 \text{ g Li}$

g. $0.01205 \text{ mol F} \times \dfrac{19.00 \text{ g}}{1 \text{ mol}} = 0.2290 \text{ g F}$

110. a. mass of 1 mol Fe = 1(55.85 g) = 55.85 g
mass of 1 mol S = 1(32.07 g) = 32.07 g
mass of 4 mol O = 4(16.00 g) = 64.00 g
molar mass of $FeSO_4$ = 151.92 g

b. mass of 1 mol Hg = 1(200.6 g) = 200.6 g
mass of 2 mol I = 2(126.9 g) = 253.8 g
molar mass of $HgI_2$ = 454.4 g

c. mass of 1 mol Sn = 1(118.7 g) = 118.7 g
mass of 2 mol O = 2(16.00 g) = 32.00 g
molar mass of $SnO_2$ = 150.7 g

d. mass of 1 mol Co = 1(58.93 g) = 58.93 g
mass of 2 mol Cl = 2(35.45 g) = 70.90 g
molar mass of $CoCl_2$ = 129.83 g

e. mass of 1 mol Cu = 1(63.55 g) = 63.55 g
mass of 2 mol N = 2(14.01 g) = 28.02 g
mass of 6 mol O = 6(16.00 g) = 96.00 g
molar mass of $Cu(NO_3)_2$ = 187.57 g

112. a. molar mass of $(NH_4)_2S$ = 68.15 g

$21.2 \text{ g} \times \dfrac{1 \text{ mol}}{68.15 \text{ g}} = 0.311 \text{ mol } (NH_4)_2S$

b. molar mass of $Ca(NO_3)_2$ = 164.1 g

Chapter 8:   Chemical Composition

$$44.3 \text{ g} \times \frac{1 \text{ mol}}{164.1 \text{ g}} = 0.270 \text{ mol Ca(NO}_3)_2$$

c. molar mass of $Cl_2O$ = 86.9 g

$$4.35 \text{ g} \times \frac{1 \text{ mol}}{86.9 \text{ g}} = 0.0501 \text{ mol Cl}_2\text{O}$$

d. 1.0 lb = 454 g; molar mass of $FeCl_3$ = 162.2

$$454 \text{ g} \times \frac{1 \text{ mol}}{162.2 \text{ g}} = 2.8 \text{ mol FeCl}_3$$

e. 1.0 kg = 1.0 × 10³ g;   molar mass of $FeCl_3$ = 162.2 g

$$1.0 \times 10^3 \text{ g} \times \frac{1 \text{ mol}}{162.2 \text{ g}} = 6.2 \text{ mol FeCl}_3$$

114. (*a*); Since there are equal mole samples, assume 1 mole of each compound. This will allow a quick determination of the greatest number of moles of oxygen in each compound and thus the greatest number of oxygen atoms (due to Avogadro's number).

$$1 \text{ mol Mg(NO}_3)_2 \times \frac{6 \text{ mol O}}{1 \text{ mol Mg(NO}_3)_2} \times \frac{6.022 \times 10^{23} \text{ O atoms}}{1 \text{ mol O}} = 3.613 \times 10^{24} \text{ O atoms}$$

$$1 \text{ mol N}_2\text{O}_5 \times \frac{5 \text{ mol O}}{1 \text{ mol N}_2\text{O}_5} \times \frac{6.022 \times 10^{23} \text{ O atoms}}{1 \text{ mol O}} = 3.011 \times 10^{24} \text{ O atoms}$$

$$1 \text{ mol FePO}_4 \times \frac{4 \text{ mol O}}{1 \text{ mol FePO}_4} \times \frac{6.022 \times 10^{23} \text{ O atoms}}{1 \text{ mol O}} = 2.409 \times 10^{24} \text{ O atoms}$$

$$1 \text{ mol BaO} \times \frac{1 \text{ mol O}}{1 \text{ mol BaO}} \times \frac{6.022 \times 10^{23} \text{ O atoms}}{1 \text{ mol O}} = 6.022 \times 10^{23} \text{ O atoms}$$

$$1 \text{ mol KC}_2\text{H}_3\text{O}_2 \times \frac{2 \text{ mol O}}{1 \text{ mol KC}_2\text{H}_3\text{O}_2} \times \frac{6.022 \times 10^{23} \text{ O atoms}}{1 \text{ mol O}} = 1.204 \times 10^{24} \text{ O atoms}.$$

116. a. molar mass of $C_6H_{12}O_6$ = 180.2 g

$$3.45 \text{ g} \times \frac{6.022 \times 10^{23} \text{ molecules}}{180.2 \text{ g}} = 1.15 \times 10^{22} \text{ molecules C}_6\text{H}_{12}\text{O}_6$$

b. $$3.45 \text{ mol} \times \frac{6.022 \times 10^{23} \text{ molecules}}{1 \text{ mol}} = 2.08 \times 10^{24} \text{ molecules C}_6\text{H}_{12}\text{O}_6$$

c. molar mass of $ICl_5$ = 304.2 g

$$25.0 \text{ g} \times \frac{6.022 \times 10^{23} \text{ molecules}}{304.2 \text{ g}} = 4.95 \times 10^{22} \text{ molecules ICl}_5$$

d. molar mass of $B_2H_6$ = 27.67 g

Chapter 8: Chemical Composition

$$1.00 \text{ g} \times \frac{6.022 \times 10^{23} \text{ molecules}}{27.67 \text{ g}} = 2.18 \times 10^{22} \text{ molecules B}_2\text{H}_6$$

e. 1.05 mmol = 0.00105 mol

$$0.00105 \text{ mol} \times \frac{6.022 \times 10^{23} \text{ formula units}}{1 \text{ mol}} = 6.32 \times 10^{20} \text{ formula units}$$

118. molar mass of CaBr$_2$ = 199.88 g; Br$^-$ ions are the anions (negatively charged ions)

$$5.00 \text{ g CaBr}_2 \times \frac{1 \text{ mol CaBr}_2}{199.88 \text{ g CaBr}_2} \times \frac{2 \text{ mol Br}^-}{1 \text{ mol CaBr}_2} \times \frac{6.022 \times 10^{23} \text{ Br}^- \text{ ions}}{1 \text{ mol Br}^-} = 3.01 \times 10^{22} \text{ anions (Br}^- \text{ ions)}$$

120. (d);

$$46.7 \text{ g N} \times \frac{1 \text{ mol N}}{14.01 \text{ g N}} = \frac{3.33 \text{ mol N}}{3.33 \text{ mol}} = 1$$

$$(100.0 - 46.7) \text{ g O} \times \frac{1 \text{ mol O}}{16.00 \text{ g O}} = \frac{3.33 \text{ mol O}}{3.33 \text{ mol}} = 1$$

The empirical formula is therefore NO.

122. $$100.0 \text{ g (NH}_4)_2\text{CO}_3 \times \frac{1 \text{ mol (NH}_4)_2\text{CO}_3}{96.09 \text{ g (NH}_4)_2\text{CO}_3} \times \frac{3 \text{ mol O}}{1 \text{ mol (NH}_4)_2\text{CO}_3} \times \frac{1 \text{ mol NaOH}}{1 \text{ mol O}} \times \frac{40.00 \text{ g NaOH}}{1 \text{ mol NaOH}} = 124.9 \text{ g NaOH}$$

124. Assume 100.0 g of hydrocortisone.

$$69.58 \text{ g C} \times \frac{1 \text{ mol C}}{12.01 \text{ g C}} \times \frac{6.022 \times 10^{23} \text{ C atoms}}{1 \text{ mol C}} = 3.489 \times 10^{24} \text{ C atoms}$$

$$3.489 \times 10^{24} \text{ C atoms} \times \frac{1 \text{ molecule hydrocortisone}}{21 \text{ atoms C}} \times \frac{1 \text{ mol hydrocortisone}}{6.022 \times 10^{23} \text{ hydrocortisone molecules}} = 0.2759 \text{ mol hydrocortisone}$$

$$\frac{100.0 \text{ g hydrocortisone}}{0.2759 \text{ mol hydrocortisone}} = 362.5 \text{ g/mol}$$

126. Assume we have 100.0 g of the compound.

$$65.95 \text{ g Ba} \times \frac{1 \text{ mol}}{137.3 \text{ g}} = 0.4803 \text{ mol Ba}$$

$$34.05 \text{ g Cl} \times \frac{1 \text{ mol}}{35.45 \text{ g}} = 0.9605 \text{ mol Cl}$$

Dividing each of these number of moles by the smaller number gives

$$\frac{0.4803 \text{ mol Ba}}{0.4803 \text{ mol}} = 1.000 \text{ mol Ba} \qquad \frac{0.9605 \text{ mol Cl}}{0.4803 \text{ mol}} = 2.000 \text{ mol Cl}$$

The empirical formula is then BaCl$_2$.

Chapter 8: Chemical Composition

128. a. $C_{63}H_{88}CoN_{14}O_{14}P$

mass of 63 mol C = 63(12.01 g) = 756.63 g

mass of 88 mol H = 88(1.008 g) = 88.704 g

mass of 1 mol Co = 58.93 g

mass of 14 mol N = 14(14.01 g) = 196.14 g

mass of 14 mol O = 14(16.00 g) = 224.00 g

mass of 1 mol P = 30.97 g

molar mass of $C_{63}H_{88}CoN_{14}O_{14}P$ = (756.63+88.704+58.93+196.14+224.00+30.97) = 1355.37 g

b. $250 \text{ mg } C_{63}H_{88}CoN_{14}O_{14}P \times \dfrac{1 \text{ g}}{1000 \text{ mg}} \times \dfrac{1 \text{ mol}}{1355.37 \text{ g}} = 1.8 \times 10^{-4} \text{ mol}$

c. $0.60 \text{ mol } C_{63}H_{88}CoN_{14}O_{14}P \times \dfrac{1355.37 \text{ g}}{1 \text{ mol}} = 810 \text{ g}$

d. $1.0 \text{ mol } C_{63}H_{88}CoN_{14}O_{14}P \times \dfrac{88 \text{ mol H}}{1 \text{ mol } C_{63}H_{88}CoN_{14}O_{14}P} \times \dfrac{6.022 \times 10^{23} \text{ H atoms}}{1 \text{ mol H}} =$

= $5.3 \times 10^{25}$ H atoms

e. $1.0 \times 10^7 \text{ molecules } C_{63}H_{88}CoN_{14}O_{14}P \times \dfrac{1 \text{ mol}}{6.022 \times 10^{23}} \times \dfrac{1355.37 \text{ g}}{1 \text{ mol}} = 2.3 \times 10^{-14} \text{ g}$

f. $1.0 \text{ molecule } C_{63}H_{88}CoN_{14}O_{14}P \times \dfrac{1 \text{ mol}}{6.022 \times 10^{23}} \times \dfrac{1355.37 \text{ g}}{1 \text{ mol}} = 2.3 \times 10^{-21} \text{ g}$

130. a. $1.0 \text{ g } CH_4O \times \dfrac{1 \text{ mol } CH_4O}{32.042 \text{ g } CH_4O} \times \dfrac{1 \text{ mol C}}{1 \text{ mol } CH_4O} \times \dfrac{6.022 \times 10^{23} \text{ C atoms}}{1 \text{ mol C}} = 1.9 \times 10^{22}$ C atoms

b. $1.0 \text{ g } CH_3CH_2OH \times \dfrac{1 \text{ mol } CH_3CH_2OH}{46.068 \text{ g } CH_3CH_2OH} \times \dfrac{2 \text{ mol C}}{1 \text{ mol } CH_3CH_2OH} \times \dfrac{6.022 \times 10^{23} \text{ C atoms}}{1 \text{ mol C}} =$

= $2.6 \times 10^{22}$ C atoms

c. $25.0 \text{ g } CO(NH_2)_2 \times \dfrac{1 \text{ mol } CO(NH_2)_2}{60.062 \text{ g } CO(NH_2)_2} \times \dfrac{2 \text{ mol N}}{1 \text{ mol } CO(NH_2)_2} \times \dfrac{6.022 \times 10^{23} \text{ N atoms}}{1 \text{ mol N}} =$

= $5.01 \times 10^{23}$ N atoms

132. molar mass of $C_9H_8O_4$ = 180.154 g

$\% \text{ C} = \dfrac{108.09 \text{ g C}}{180.154 \text{ g}} \times 100 = 60.00\%$ C

$\% \text{ H} = \dfrac{8.064 \text{ g H}}{180.154 \text{ g}} \times 100 = 4.476\%$ H

$$\% \text{ O} = \frac{64.00 \text{ g O}}{180.154 \text{ g}} \times 100 = 35.53\% \text{ O}$$

134. Consider 100.0 g of the compound so that percentages become masses.

$$40.0 \text{ g C} \times \frac{1 \text{ mol C}}{12.01 \text{ g}} = 3.33 \text{ mol C}$$

$$6.70 \text{ g H} \times \frac{1 \text{ mol H}}{1.008 \text{ g}} = 6.65 \text{ mol H}$$

$$53.3 \text{ g O} \times \frac{1 \text{ mol}}{16.00 \text{ g}} = 3.33 \text{ mol O}$$

Dividing each number of moles by the smaller number of moles gives

$$\frac{3.33 \text{ mol C}}{3.33 \text{ mol}} = 1.00 \text{ mol C}$$

$$\frac{6.65 \text{ mol H}}{3.33 \text{ mol}} = 2.00 \text{ mol H}$$

$$\frac{3.33 \text{ mol O}}{3.33 \text{ mol}} = 1.000 \text{ mol O}$$

The empirical formula is therefore $CH_2O$ (with a molar mass of 30.026 g/mol). The molar mass of the compound is 180.1 g/mol, thus the empirical formula mass goes into the molecular formula mass 6 times (180.1/30.026 = 6). The molecular formula is $6 \times (CH_2O) = C_6H_{12}O_6$.

# CHAPTER 9

# Chemical Quantities

2. The coefficients of this balanced chemical equation indicate the relative numbers of molecules (or moles) of each reactant that combine, as well as the number of molecules (or moles) of the product formed.

4. (e); Subscripts cannot be changed to balance the equation or else the identities of the reactants and/or product are changed. The balanced equation is: $N_2(g) + 3H_2(g) \rightarrow 2NH_3(g)$. Nitrogen and hydrogen will react regardless of how much is initially present (just one might be used up before the other).

6. a. $3MnO_2(s) + 4Al(s) \rightarrow 3Mn(s) + 2Al_2O_3(s)$

   Three formula units (or three moles) of manganese(IV) oxide react with four atoms (or four moles) of aluminum, producing three atoms (or three moles) of manganese and two formula units (or two moles) of aluminum oxide.

   b. $B_2O_3(s) + 3CaF_2(s) \rightarrow 2BF_3(g) + 3CaO(s)$

   One molecule (or one mole) of diboron trioxide reacts with three formula units (or three moles) of calcium fluoride, producing two molecules (or two moles) of boron trifluoride and three formula units (three moles) of calcium oxide.

   c. $3NO_2(g) + H_2O(l) \rightarrow 2HNO_3(aq) + NO(g)$

   Three molecules (or three moles) of nitrogen dioxide react with one molecule (or one mole) of water, producing two molecules (or two moles) of nitric acid and one molecule (or one mole) of nitrogen monoxide.

   d. $C_6H_6(g) + 3H_2(g) \rightarrow C_6H_{12}(g)$

   One molecule (or one mole) of benzene ($C_6H_6$) reacts with three molecules (or three moles) of hydrogen gas, producing one molecule (or one mole) of cyclohexane ($C_6H_{12}$).

8. Balanced chemical equations tell us in what molar ratios substances combine to form products, not in what mass proportions they combine. How could a total of 3 g of reactants produce 2 g of product?

10. The balanced equation is: $2C_7H_{14} + 21O_2 \rightarrow 14CO_2 + 14H_2O$

    For converting from a given number of moles of heptene to the number of moles of oxygen required, the mole ratio is

    $$\left( \frac{21 \text{ mol } O_2}{2 \text{ mol } C_7H_{14}} \right)$$

    For a given number of moles of heptene reacting completely, the mole ratios used to calculate the number of moles of each product are

Chapter 9: Chemical Quantities

For $CO_2$: $\left(\dfrac{14 \text{ mol } CO_2}{2 \text{ mol } C_7H_{14}}\right)$  For $H_2O$: $\left(\dfrac{14 \text{ mol } H_2O}{2 \text{ mol } C_7H_{14}}\right)$

12. Before doing the calculations, the equations must be *balanced*.

   a.  $2FeO(s) + C(s) \rightarrow 2Fe(l) + CO_2(g)$

   $0.125 \text{ mol FeO} \times \dfrac{2 \text{ mol Fe}}{2 \text{ mol FeO}} = 0.125 \text{ mol Fe}$

   $0.125 \text{ mol FeO} \times \dfrac{1 \text{ mol } CO_2}{2 \text{ mol FeO}} = 0.0625 \text{ mol } CO_2$

   b.  $Cl_2(g) + 2KI(aq) \rightarrow 2KCl(aq) + I_2(s)$

   $0.125 \text{ mol KI} \times \dfrac{2 \text{ mol KCl}}{2 \text{ mol KI}} = 0.125 \text{ mol KCl}$

   $0.125 \text{ mol KI} \times \dfrac{1 \text{ mol } I_2}{2 \text{ mol KI}} = 0.0625 \text{ mol } I_2$

   c.  $2Na_2B_4O_7(s) + 2H_2SO_4(aq) + 10H_2O(l) \rightarrow 8H_3BO_3(s) + 2Na_2SO_4(aq)$

   $0.125 \text{ mol } Na_2B_4O_7 \times \dfrac{8 \text{ mol } H_3BO_3}{2 \text{ mol } Na_2B_4O_7} = 0.500 \text{ mol } H_3BO_3$

   $0.125 \text{ mol } Na_2B_4O_7 \times \dfrac{2 \text{ mol } Na_2SO_4}{2 \text{ mol } Na_2B_4O_7} = 0.125 \text{ mol } Na_2SO_4$

   d.  $CaC_2(s) + 2H_2O(l) \rightarrow Ca(OH)_2(s) + C_2H_2(g)$

   $0.125 \text{ mol } CaC_2 \times \dfrac{1 \text{ mol } Ca(OH)_2}{1 \text{ mol } CaC_2} = 0.125 \text{ mol } Ca(OH)_2$

   $0.125 \text{ mol } CaC_2 \times \dfrac{1 \text{ mol } C_2H_2}{1 \text{ mol } CaC_2} = 0.125 \text{ mol } C_2H_2$

14. a.  $NH_3(g) + HCl(g) \rightarrow NH_4Cl(s)$
   molar mass: $NH_4Cl$, 53.492 g

   $0.50 \text{ mol } NH_3 \times \dfrac{1 \text{ mol } NH_4Cl}{1 \text{ mol } NH_3} = 0.50 \text{ mol } NH_4Cl$

   $0.50 \text{ mol } NH_4Cl \times \dfrac{53.492 \text{ g } NH_4Cl}{1 \text{ mol } NH_3} = 27 \text{ g } NH_4Cl$

   b.  $CH_4(g) + 4S(s) \rightarrow CS_2(l) + 2H_2S(g)$
   molar masses: $CS_2$, 76.15 g; $H_2S$, 34.086 g

   $0.50 \text{ mol S} \times \dfrac{1 \text{ mol } CS_2}{4 \text{ mol S}} = 0.125 \text{ mol } CS_2$

Chapter 9: Chemical Quantities

$$0.125 \text{ mol } CS_2 \times \frac{76.15 \text{ g } CS_2}{1 \text{ mol } CS_2} = 9.5 \text{ g } CS_2$$

$$0.50 \text{ mol } S \times \frac{2 \text{ mol } H_2S}{4 \text{ mol } S} = 0.25 \text{ mol } H_2S$$

$$0.25 \text{ mol } H_2S \times \frac{34.086 \text{ g } H_2S}{1 \text{ mol } H_2S} = 8.5 \text{ g } H_2S$$

c. $PCl_3(l) + 3H_2O(l) \rightarrow H_3PO_3(aq) + 3HCl(aq)$

molar masses: $H_3PO_3$, 81.994 g; HCl, 36.458 g

$$0.50 \text{ mol } PCl_3 \times \frac{1 \text{ mol } H_3PO_3}{1 \text{ mol } PCl_3} = 0.50 \text{ mol } H_3PO_3$$

$$0.50 \text{ mol } H_3PO_3 \times \frac{81.994 \text{ g } H_3PO_3}{1 \text{ mol } H_3PO_3} = 41 \text{ g } H_3PO_3$$

$$0.50 \text{ mol } PCl_3 \times \frac{3 \text{ mol } HCl}{1 \text{ mol } PCl_3} = 1.50 \text{ mol } HCl$$

$$1.50 \text{ mol } HCl \times \frac{36.458 \text{ g } HCl}{1 \text{ mol } HCl} = 55 \text{ g } HCl$$

d. $NaOH(s) + CO_2(g) \rightarrow NaHCO_3(s)$

molar masses: $NaHCO_3$, 84.008 g

$$0.50 \text{ mol } NaOH \times \frac{1 \text{ mol } NaHCO_3}{1 \text{ mol } NaOH} = 0.50 \text{ mol } NaHCO_3$$

$$0.50 \text{ mol } NaHCO_3 \times \frac{84.008 \text{ g } NaHCO_3}{1 \text{ mol } NaHCO_3} = 42 \text{ g } NaHCO_3$$

16. Before doing the calculations, the equations must be *balanced*.

   a. $4KO_2(s) + 2H_2O(l) \rightarrow 3O_2(g) + 4KOH(s)$

   $$0.625 \text{ mol } KOH \times \frac{3 \text{ mol } O_2}{4 \text{ mol } KOH} = 0.469 \text{ mol } O_2$$

   b. $SeO_2(g) + 2H_2Se(g) \rightarrow 3Se(s) + 2H_2O(g)$

   $$0.625 \text{ mol } H_2O \times \frac{3 \text{ mol } Se}{2 \text{ mol } H_2O} = 0.938 \text{ mol } Se$$

   c. $2CH_3CH_2OH(l) + O_2(g) \rightarrow 2CH_3CHO(aq) + 2H_2O(l)$

   $$0.625 \text{ mol } H_2O \times \frac{2 \text{ mol } CH_3CHO}{2 \text{ mol } H_2O} = 0.625 \text{ mol } CH_3CHO$$

Chapter 9: Chemical Quantities

d. $Fe_2O_3(s) + 2Al(s) \rightarrow 2Fe(l) + Al_2O_3(s)$

$0.625 \text{ mol } Al_2O_3 \times \dfrac{2 \text{ mol Fe}}{1 \text{ mol } Al_2O_3} = 1.25 \text{ mol Fe}$

18. Stoichiometry is the process of using a chemical equation to calculate the relative masses (or moles) of reactants and products involved in a reaction.

20. a. molar mass Ag = 107.9 g

$2.01 \times 10^{-2} \text{ g Ag} \times \dfrac{1 \text{ mol Ag}}{107.9 \text{ g Ag}} = 1.86 \times 10^{-4} \text{ mol Ag}$

b. molar mass $(NH_4)_2S$ = 68.154 g

$45.2 \text{ mg } (NH_4)_2S \times \dfrac{1 \text{ g}}{1000 \text{ mg}} \times \dfrac{1 \text{ mol } (NH_4)_2S}{68.154 \text{ g } (NH_4)_2S} = 6.63 \times 10^{-4} \text{ mol } (NH_4)_2S$

c. molar mass U = 238.00 g

$61.7 \text{ μg U} \times \dfrac{1 \text{ g}}{10^6 \text{ μg}} \times \dfrac{1 \text{ mol U}}{238.00 \text{ g U}} = 2.59 \times 10^{-7} \text{ mol U}$

d. molar mass $SO_2$ = 64.07 g

$5.23 \text{ kg } SO_2 \times \dfrac{1000 \text{ g}}{1 \text{ kg}} \times \dfrac{1 \text{ mol } SO_2}{64.07 \text{ g } SO_2} = 81.6 \text{ mol } SO_2$

e. molar mass $Fe(NO_3)_3$ = 241.88 g

$272 \text{ g } Fe(NO_3)_3 \times \dfrac{1 \text{ mol } Fe(NO_3)_3}{241.88 \text{ g } Fe(NO_3)_3} = 1.12 \text{ mol } Fe(NO_3)_3$

22. a. molar mass of $K_3N$ = 131.31 g

$0.341 \text{ mol } K_3N \times \dfrac{131.31 \text{ g } K_3N}{1 \text{ mol } K_3N} = 44.8 \text{ g } K_3N$

b. molar mass of Ne = 20.18 g; 2.62 millimol = 0.00262 mol

$0.00262 \text{ mol Ne} \times \dfrac{20.18 \text{ g Ne}}{1 \text{ mol Ne}} = 0.0529 \text{ g Ne}$

c. molar mass of MnO = 70.94 g

$0.00449 \text{ mol MnO} \times \dfrac{70.94 \text{ g MnO}}{1 \text{ mol MnO}} = 0.319 \text{ g MnO}$

d. molar mass of $SiO_2$ = 60.09 g

$7.18 \times 10^5 \text{ mol } SiO_2 \times \dfrac{60.09 \text{ g } SiO_2}{1 \text{ mol } SiO_2} = 4.31 \times 10^7 \text{ g } SiO_2$

Chapter 9: Chemical Quantities

e. molar mass of FePO$_4$ = 150.82 g

$$0.000121 \text{ mol FePO}_4 \times \frac{150.82 \text{ g FePO}_4}{1 \text{ mol FePO}_4} = 0.0182 \text{ g FePO}_4$$

24. Before any calculations are done, the equations must be *balanced*.

   a. $2Al(s) + 3Br_2(l) \rightarrow 2AlBr_3(s)$

   molar mass Al = 26.98 g

   $$0.557 \text{ g Al} \times \frac{1 \text{ mol Al}}{26.98 \text{ g Al}} \times \frac{3 \text{ mol Br}_2}{2 \text{ mol Al}} = 0.0310 \text{ mol Br}_2$$

   b. $Hg(s) + 2HClO_4(aq) \rightarrow Hg(ClO_4)_2(aq) + H_2(g)$

   molar mass Hg = 200.6 g

   $$0.557 \text{ g Hg} \times \frac{1 \text{ mol Hg}}{200.6 \text{ g}} \times \frac{2 \text{ mol HClO}_4}{1 \text{ mol Hg}} = 0.00555 \text{ mol HClO}_4$$

   c. $3K(s) + P(s) \rightarrow K_3P(s)$

   molar mass K = 39.10 g

   $$0.557 \text{ g K} \times \frac{1 \text{ mol K}}{39.10 \text{ g K}} \times \frac{1 \text{ mol P}}{3 \text{ mol K}} = 0.00475 \text{ mol P}$$

   d. $CH_4(g) + 4Cl_2(g) \rightarrow CCl_4(l) + 4HCl(g)$

   molar mass CH$_4$ = 16.04 g

   $$0.557 \text{ g CH}_4 \times \frac{1 \text{ mol CH}_4}{16.04 \text{ g CH}_4} \times \frac{4 \text{ mol Cl}_2}{1 \text{ mol CH}_4} = 0.139 \text{ mol Cl}_2$$

26. a. molar masses: NI$_3$, 394.71 g; IF, 145.9 g; BF$_3$, 67.81 g

   $$30.0 \text{ g NI}_3 \times \frac{1 \text{ mol NI}_3}{394.71 \text{ g NI}_3} = 0.0760 \text{ mol NI}_3$$

   $$0.0760 \text{ mol NI}_3 \times \frac{3 \text{ mol IF}}{1 \text{ mol NI}_3} = 0.228 \text{ mol IF}$$

   $$0.228 \text{ mol IF} \times \frac{145.9 \text{ g IF}}{1 \text{ mol IF}} = 33.3 \text{ g IF}$$

   b. $$0.0760 \text{ mol NI}_3 \times \frac{1 \text{ mol BF}_3}{1 \text{ mol NI}_3} = 0.0760 \text{ mol BF}_3$$

   $$0.0760 \text{ mol BF}_3 \times \frac{67.81 \text{ g BF}_3}{1 \text{ mol BF}_3} = 5.15 \text{ g BF}_3$$

28. The balanced equation for the reaction is:

   $CaC_2(s) + 2H_2O(l) \rightarrow C_2H_2(g) + Ca(OH)_2(s)$

molar masses: CaC$_2$, 64.10 g; C$_2$H$_2$, 26.04 g

$$3.75 \text{ g CaC}_2 \times \frac{1 \text{ mol CaC}_2}{64.10 \text{ g CaC}_2} = 0.0585 \text{ mol CaC}_2$$

$$0.0585 \text{ mol CaC}_2 \times \frac{1 \text{ mol C}_2\text{H}_2}{1 \text{ mol CaC}_2} = 0.0585 \text{ mol C}_2\text{H}_2$$

$$0.0585 \text{ mol C}_2\text{H}_2 \times \frac{26.04 \text{ g C}_2\text{H}_2}{1 \text{ mol C}_2\text{H}_2} = 1.52 \text{ g C}_2\text{H}_2$$

30.  2NaHCO$_3$(s) → Na$_2$CO$_3$(s) + H$_2$O(g) + CO$_2$(g)

molar masses: NaHCO$_3$, 84.01 g; Na$_2$CO$_3$, 106.0 g

$$1.52 \text{ g NaHCO}_3 \times \frac{1 \text{ mol NaHCO}_3}{84.01 \text{ g NaHCO}_3} = 0.01809 \text{ mol NaHCO}_3$$

$$0.01809 \text{ mol NaHCO}_3 \times \frac{1 \text{ mol Na}_2\text{CO}_3}{2 \text{ mol NaHCO}_3} = 0.009047 \text{ mol Na}_2\text{CO}_3$$

$$0.009047 \text{ mol Na}_2\text{CO}_3 \times \frac{106.0 \text{ g Na}_2\text{CO}_3}{1 \text{ mol Na}_2\text{CO}_3} = 0.959 \text{ g Na}_2\text{CO}_3$$

32.  C$_6$H$_{12}$O$_6$(aq) → 2C$_2$H$_5$OH(aq) + 2CO$_2$(g)

molar masses: C$_6$H$_{12}$O$_6$, 180.2 g; C$_2$H$_5$OH, 46.07 g

$$5.25 \text{ g C}_6\text{H}_{12}\text{O}_6 \times \frac{1 \text{ mol C}_6\text{H}_{12}\text{O}_6}{180.2 \text{ g C}_6\text{H}_{12}\text{O}_6} = 0.02913 \text{ mol C}_6\text{H}_{12}\text{O}_6$$

$$0.02913 \text{ mol C}_6\text{H}_{12}\text{O}_6 \times \frac{2 \text{ mol C}_2\text{H}_5\text{OH}}{1 \text{ mol C}_6\text{H}_{12}\text{O}_6} = 0.5826 \text{ mol C}_2\text{H}_5\text{OH}$$

$$0.5286 \text{ mol C}_2\text{H}_5\text{OH} \times \frac{46.07 \text{ g C}_2\text{H}_5\text{OH}}{1 \text{ mol C}_2\text{H}_5\text{OH}} = 2.68 \text{ g ethyl alcohol}$$

34.  NH$_4$Cl(s) + NaOH(s) → NH$_3$(g) + NaCl(s) + H$_2$O(g)

molar masses: NH$_4$Cl, 53.49 g; NH$_3$, 17.03 g

$$1.39 \text{ g NH}_4\text{Cl} \times \frac{1 \text{ mol NH}_4\text{Cl}}{53.49 \text{ g NH}_4\text{Cl}} = 0.02599 \text{ mol NH}_4\text{Cl}$$

$$0.02599 \text{ mol NH}_4\text{Cl} \times \frac{1 \text{ mol NH}_3}{1 \text{ mol NH}_4\text{Cl}} = 0.02599 \text{ mol NH}_3$$

$$0.02599 \text{ mol NH}_3 \times \frac{17.03 \text{ g NH}_3}{1 \text{ mol NH}_3} = 0.443 \text{ g NH}_3$$

## Chapter 9: Chemical Quantities

**36.** $4HgS(s) + 4CaO(s) \rightarrow 4Hg(l) + 3CaS(s) + CaSO_4(s)$

molar masses: HgS, 232.7 g Hg, 200.6 g; 10.0 kg = $1.00 \times 10^4$ g

$$1.00 \times 10^4 \text{ g HgS} \times \frac{1 \text{ mol HgS}}{232.7 \text{ g HgS}} = 42.97 \text{ mol HgS}$$

$$42.97 \text{ mol HgS} \times \frac{4 \text{ mol Hg}}{4 \text{ mol HgS}} = 42.97 \text{ mol Hg}$$

$$42.97 \text{ mol Hg} \times \frac{200.6 \text{ g Hg}}{1 \text{ mol Hg}} = 8.62 \times 10^3 \text{ g Hg} = 8.62 \text{ kg Hg}$$

**38.** $C_{12}H_{22}O_{11}(s) \rightarrow 12C(s) + 11H_2O(g)$

molar masses: $C_{12}H_{22}O_{11}$, 342.3 g; C, 12.01

$$1.19 \text{ g } C_{12}H_{22}O_{11} \times \frac{1 \text{ mol } C_{12}H_{22}O_{11}}{342.3 \text{ g } C_{12}H_{22}O_{11}} = 3.476 \times 10^{-3} \text{ mol } C_{12}H_{22}O_{11}$$

$$3.476 \times 10^{-3} \text{ mol } C_{12}H_{22}O_{11} \times \frac{12 \text{ mol C}}{1 \text{ mol } C_{12}H_{22}O_{11}} = 0.04172 \text{ mol C}$$

$$0.04172 \text{ mol C} \times \frac{12.01 \text{ g C}}{1 \text{ mol C}} = 0.501 \text{ g C}$$

**40.** The balanced equation is:

$2C_8H_{18} + 25O_2 \rightarrow 16CO_2 + 18H_2O$

molar masses: $C_8H_{18}$, 114.22 g; $CO_2$, 44.01 g     1 lb of $CO_2$ = 453.59 g $CO_2$

$$453.59 \text{ g } CO_2 \times \frac{1 \text{ mol } CO_2}{44.01 \text{ g } CO_2} = 10.31 \text{ mol } CO_2$$

From the balanced chemical equation, we can calculate the number of moles and number of grams of pure octane that would be required to produce 10.31 mol $CO_2$.

$$10.31 \text{ mol } CO_2 \times \frac{2 \text{ mol } C_8H_{18}}{16 \text{ mol } CO_2} = 1.288 \text{ mol } C_8H_{18}$$

$$1.288 \text{ mol } C_8H_{18} \times \frac{114.22 \text{ g } C_8H_{18}}{1 \text{ mol } C_8H_{18}} = 147.2 \text{ g } C_8H_{18}$$

From the density of $C_8H_{18}$ we can calculate the volume of 147.2 g $C_8H_{18}$.

$$147.2 \text{ g } C_8H_{18} \times \frac{1 \text{ mL } C_8H_{18}}{0.75 \text{ g } C_8H_{18}} = 196.3 \text{ mL } C_8H_{18} \text{ } (2.0 \times 10^2 \text{ mL to two significant figures})$$

From the preceding, we know that to travel 1 mile, we need approximately 200 mL of octane

$$\frac{1 \text{ mi}}{196.3 \text{ mL}} \times \frac{1000 \text{ mL}}{1 \text{ L}} \times \frac{3.7854 \text{ L}}{1 \text{ gal}} = \text{approximately 19 mi/gal}$$

Chapter 9: Chemical Quantities

42. To determine the limiting reactant, first calculate the number of moles of each reactant present (which is dependent on both the starting mass and molar mass of each reactant). Then determine how these numbers of moles correspond to the stoichiometric ratio indicated by the balanced chemical equation for the reaction (the *Change* row in a BCA table). For example, to determine the limiting reactant between a reaction using 5 moles of $N_2$ and 9 moles of $H_2$, setup a BCA table with the balanced chemical equation:

$$N_2 + 3H_2 \rightarrow 2NH_3$$

| | $N_2$ | $3H_2$ | $2NH_3$ |
|---|---|---|---|
| Before: | 5 | 9 | 0 |
| Change: | −3 | −9 | +6 |
| After: | 2 | 0 | 6 |

Since $H_2$ completely runs out and no negative amounts of substance were left afterwards, $H_2$ is the limiting reactant.

44. (e); The limiting reactant cannot be determined since the starting amounts are not given.

46. a. $S(s) + 2H_2SO_4(aq) \rightarrow 3SO_2(g) + 2H_2O(l)$

Molar masses: S, 32.07 g; $H_2SO_4$, 98.09 g; $SO_2$, 64.07 g; $H_2O$, 18.02 g

$$5.00 \text{ g S} \times \frac{1 \text{ mol}}{32.07 \text{ g}} = 0.1559 \text{ mol S}$$

$$5.00 \text{ g H}_2\text{SO}_4 \times \frac{1 \text{ mol}}{98.09 \text{ g}} = 0.05097 \text{ mol H}_2\text{SO}_4$$

According to the balanced chemical equation, we would need twice as much sulfuric acid as sulfur for complete reaction of both reactants. We clearly have much less sulfuric acid present than sulfur: sulfuric acid is the limiting reactant. The calculation of the masses of products produced is based on the number of moles of the sulfuric acid.

$$0.05097 \text{ mol H}_2\text{SO}_4 \times \frac{3 \text{ mol SO}_2}{2 \text{ mol H}_2\text{SO}_4} \times \frac{64.07 \text{ g SO}_2}{1 \text{ mol SO}_2} = 4.90 \text{ g SO}_2$$

$$0.05097 \text{ mol H}_2\text{SO}_4 \times \frac{2 \text{ mol H}_2\text{O}}{2 \text{ mol H}_2\text{SO}_4} \times \frac{18.02 \text{ g H}_2\text{O}}{1 \text{ mol H}_2\text{O}} = 0.918 \text{ g H}_2\text{O}$$

b. $MnO_2(s) + 2H_2SO_4(aq) \rightarrow Mn(SO_4)_2 + 2H_2O(l)$

molar masses: $MnO_2$, 86.94 g; $H_2SO_4$ 98.09 g; $Mn(SO_4)_2$, 247.1 g; $H_2O$, 18.02 g

$$5.00 \text{ g MnO}_2 \times \frac{1 \text{ mol}}{86.94 \text{ g}} = 0.05751 \text{ mol MnO}_2$$

$$5.00 \text{ g H}_2\text{SO}_4 \times \frac{1 \text{ mol}}{98.09 \text{ g}} = 0.05097 \text{ mol H}_2\text{SO}_4$$

According to the balanced chemical equation, we would need twice as much sulfuric acid as manganese(IV) oxide for complete reaction of both reactants. We do not have this much sulfuric acid, so sulfuric acid must be the limiting reactant. The amount of each product produced will be based on the sulfuric acid reacting completely.

$$0.05097 \text{ mol H}_2\text{SO}_4 \times \frac{1 \text{ mol Mn(SO}_4)_2}{2 \text{ mol H}_2\text{SO}_4} \times \frac{247.1 \text{ g Mn(SO}_4)_2}{1 \text{ mol Mn(SO}_4)_2} = 6.30 \text{ g Mn(SO}_4)_2$$

$$0.05097 \text{ mol H}_2\text{SO}_4 \times \frac{2 \text{ mol H}_2\text{O}}{2 \text{ mol H}_2\text{SO}_4} \times \frac{18.02 \text{ g H}_2\text{O}}{1 \text{ mol H}_2\text{O}} = 0.918 \text{ g H}_2\text{O}$$

c.  $2H_2S(g) + 3O_2(g) \rightarrow 2SO_2(g) + 2H_2O(l)$

Molar masses: $H_2S$, 34.09 g; $O_2$, 32.00 g; $SO_2$, 64.07 g; $H_2O$, 18.02 g

$$5.00 \text{ g H}_2\text{S} \times \frac{1 \text{ mol}}{34.09 \text{ g}} = 0.1467 \text{ mol H}_2\text{S}$$

$$5.00 \text{ g O}_2 \times \frac{1 \text{ mol}}{32.00 \text{ g}} = 0.1563 \text{ mol O}_2$$

According to the balanced equation, we would need 1.5 times as much $O_2$ as $H_2S$ for complete reaction of both reactants. We don't have that much $O_2$, so $O_2$ must be the limiting reactant that will control the masses of each product produced.

$$0.1563 \text{ mol O}_2 \times \frac{2 \text{ mol SO}_2}{3 \text{ mol O}_2} \times \frac{64.07 \text{ g SO}_2}{1 \text{ mol SO}_2} = 6.67 \text{ g SO}_2$$

$$0.1563 \text{ mol O}_2 \times \frac{2 \text{ mol H}_2\text{O}}{3 \text{ mol O}_2} \times \frac{18.02 \text{ g H}_2\text{O}}{1 \text{ mol H}_2\text{O}} = 1.88 \text{ g H}_2\text{O}$$

d.  $3AgNO_3(aq) + Al(s) \rightarrow 3Ag(s) + Al(NO_3)_3(aq)$

Molar masses: $AgNO_3$, 169.9 g; Al, 26.98 g; Ag, 107.9 g; $Al(NO_3)_3$, 213.0 g

$$5.00 \text{ g AgNO}_3 \times \frac{1 \text{ mol}}{169.9 \text{ g}} = 0.02943 \text{ mol AgNO}_3$$

$$5.00 \text{ g Al} \times \frac{1 \text{ mol}}{26.98 \text{ g}} = 0.1853 \text{ mol Al}$$

According to the balanced chemical equation, we would need three moles of $AgNO_3$ for every mole of Al for complete reaction of both reactants. We in fact have fewer moles of $AgNO_3$ than aluminum, so $AgNO_3$ must be the limiting reactant. The amount of product produced is calculated from the number of moles of the limiting reactant present:

$$0.02943 \text{ mol AgNO}_3 \times \frac{3 \text{ mol Ag}}{3 \text{ mol AgNO}_3} \times \frac{107.9 \text{ g Ag}}{1 \text{ mol Ag}} = 3.18 \text{ g Ag}$$

$$0.02943 \text{ mol AgNO}_3 \times \frac{1 \text{ mol Al(NO}_3)_3}{3 \text{ mol AgNO}_3} \times \frac{213.0 \text{ g Al(NO}_3)_3}{1 \text{ mol Al(NO}_3)_3} = 2.09 \text{ g}$$

48. a. $CS_2(l) + 3O_2(g) \rightarrow CO_2(g) + 2SO_2(g)$

Molar masses: $CS_2$, 76.15 g; $O_2$, 32.00 g; $CO_2$, 44.01 g

$$1.00 \text{ g } CS_2 \times \frac{1 \text{ mol}}{76.15 \text{ g}} = 0.01313 \text{ mol } CS_2$$

$$1.00 \text{ g } O_2 \times \frac{1 \text{ mol}}{32.00 \text{ g}} = 0.03125 \text{ mol } O_2$$

From the balanced chemical equation, we would need three times as much oxygen as carbon disulfide for complete reaction of both reactants. We do not have this much oxygen, and so oxygen must be the limiting reactant.

$$0.03125 \text{ mol } O_2 \times \frac{1 \text{ mol } CO_2}{3 \text{ mol } O_2} \times \frac{44.01 \text{ g } CO_2}{1 \text{ mol } CO_2} = 0.458 \text{ g } CO_2$$

b. $2NH_3(g) + CO_2(g) \rightarrow CN_2H_4O(s) + H_2O(l)$

Molar masses: $NH_3$, 17.03 g; $CO_2$, 44.01 g; $H_2O$, 18.02 g

$$1.00 \text{ g } NH_3 \times \frac{1 \text{ mol}}{17.03 \text{ g}} = 0.05872 \text{ mol } NH_3$$

$$1.00 \text{ g } CO_2 \times \frac{1 \text{ mol}}{44.01 \text{ g}} = 0.02272 \text{ mol } CO_2$$

The balanced chemical equation tells us that we would need twice as many moles of ammonia as carbon dioxide for complete reaction of both reactants. We have *more* than this amount of ammonia present, so the reaction will be limited by the amount of carbon dioxide present.

$$0.02272 \text{ mol } CO_2 \times \frac{1 \text{ mol } H_2O}{1 \text{ mol } CO_2} \times \frac{18.02 \text{ g } H_2O}{1 \text{ mol } H_2O} = 0.409 \text{ g } H_2O$$

c. $H_2(g) + MnO_2(s) \rightarrow MnO(s) + H_2O(l)$

Molar masses: $H_2$, 2.016 g; $MnO_2$, 86.94 g; $H_2O$, 18.02 g

$$1.00 \text{ g } H_2 \times \frac{1 \text{ mol}}{2.016 \text{ g}} = 0.496 \text{ mol } H_2$$

$$1.00 \text{ g } MnO_2 \times \frac{1 \text{ mol}}{86.94 \text{ g}} = 0.0115 \text{ mol } MnO_2$$

Because the coefficients of both reactants in the balanced chemical equation are the same, we would need equal amounts of both reactants for complete reaction. Therefore, manganese(IV) oxide must be the limiting reactant and controls the amount of product obtained.

$$0.0115 \text{ mol } MnO_2 \times \frac{1 \text{ mol } H_2O}{1 \text{ mol } MnO_2} \times \frac{18.02 \text{ g } H_2O}{1 \text{ mol } H_2O} = 0.207 \text{ g } H_2O$$

d. $I_2(s) + Cl_2(g) \rightarrow 2ICl(g)$

Molar masses: $I_2$, 253.8 g; $Cl_2$, 70.90 g; $ICl$, 162.35 g

$$1.00 \text{ g I}_2 \times \frac{1 \text{ mol}}{253.8 \text{ g}} = 0.00394 \text{ mol I}_2$$

$$1.00 \text{ g Cl}_2 \times \frac{1 \text{ mol}}{70.90 \text{ g}} = 0.0141 \text{ mol Cl}_2$$

From the balanced chemical equation, we would need equal amounts of $I_2$ and $Cl_2$ for complete reaction of both reactants. As we have much less iodine than chlorine, iodine must be the limiting reactant.

$$0.00394 \text{ mol I}_2 \times \frac{2 \text{ mol ICl}}{1 \text{ mol I}_2} \times \frac{162.35 \text{ g ICl}}{1 \text{ mol ICl}} = 1.28 \text{ g ICl}$$

50. a. $2Al(s) + 6HCl(aq) \rightarrow 2AlCl_3(aq) + 3H_2(g)$

   HCl is the limiting reactant; 18.3 g $AlCl_3$; 0.415 g $H_2$

   b. $2NaOH(aq) + CO_2(g) \rightarrow Na_2CO_3(aq) + H_2O(l)$

   NaOH is the limiting reactant; 19.9 g $Na_2CO_3$; 3.38 g $H_2O$

   c. $Pb(NO_3)_2(aq) + 2HCl(aq) \rightarrow PbCl_2(s) + 2HNO_3(aq)$

   $Pb(NO_3)_2$ is the limiting reactant; 12.6 g $PbCl_2$; 5.71 g $HNO_3$

   d. $2K(s) + I_2(s) \rightarrow 2KI(s)$

   $I_2$ is the limiting reactant; 19.6 g KI

52. $CuO(s) + H_2SO_4(aq) \rightarrow CuSO_4(aq) + H_2O(l)$

   molar masses: CuO, 79.55 g; $H_2SO_4$, 98.09 g

   $$2.49 \text{ g CuO} \times \frac{1 \text{ mol CuO}}{79.55 \text{ g CuO}} = 0.0313 \text{ mol CuO}$$

   $$5.05 \text{ g H}_2\text{SO}_4 \times \frac{1 \text{ mol H}_2\text{SO}_4}{98.09 \text{ g H}_2\text{SO}_4} = 0.0515 \text{ mol H}_2\text{SO}_4$$

   Since the reaction is of 1:1 stoichiometry, CuO must be the limiting reactant since it is present in the lesser amount on a molar basis.

54. $4Fe(s) + 3O_2(g) \rightarrow 2Fe_2O_3(s)$

   Molar masses: Fe, 55.85 g; $Fe_2O_3$, 159.7 g

   $$1.25 \text{ g Fe} \times \frac{1 \text{ mol}}{55.85 \text{ g}} = 0.0224 \text{ mol Fe present}$$

   Calculate how many mol of $O_2$ are required to react with this amount of Fe

   $$0.0224 \text{ mol Fe} \times \frac{3 \text{ mol O}_2}{4 \text{ mol Fe}} = 0.0168 \text{ mol O}_2$$

   Because we have more $O_2$ than this, Fe must be the limiting reactant.

Chapter 9: Chemical Quantities

$$0.0224 \text{ mol Fe} \times \frac{2 \text{ mol Fe}_2\text{O}_3}{4 \text{ mol Fe}} \times \frac{159.7 \text{ g Fe}_2\text{O}_3}{1 \text{ mol Fe}_2\text{O}_3} = 1.79 \text{ g Fe}_2\text{O}_3$$

56. $\text{CaCl}_2(aq) + \text{Na}_2\text{SO}_4(aq) \rightarrow \text{CaSO}_4(s) + 2\text{NaCl}(aq)$

    molar masses: $\text{CaCl}_2$, 110.98 g; $\text{Na}_2\text{SO}_4$, 142.05 g

    $$5.21 \text{ g CaCl}_2 \times \frac{1 \text{ mol CaCl}_2}{110.98 \text{ g CaCl}_2} = 0.0469 \text{ mol CaCl}_2 = 0.0469 \text{ mol Ca}^{2+} \text{ ion}$$

    $$4.95 \text{ g Na}_2\text{SO}_4 \times \frac{1 \text{ mol Na}_2\text{SO}_4}{142.05 \text{ g Na}_2\text{SO}_4} = 0.0348 \text{ mol Na}_2\text{SO}_4 = 0.0348 \text{ mol SO}_4^{2-} \text{ ion}$$

    Because the balanced chemical equation indicates a 1:1 stoichiometry for the reaction, there is not nearly enough sulfate ion present (0.0348 mol) to precipitate the amount of calcium ion in the sample (0.0469 mol). Sodium sulfate (sulfate ion) is the limiting reactant. Calcium chloride (calcium ion) is present in excess.

58. $\text{SiO}_2(s) + 3\text{C}(s) \rightarrow 2\text{CO}(g) + \text{SiC}(s)$

    molar masses: $\text{SiO}_2$, 60.09 g; SiC, 40.10 g; 1.0 kg = $1.0 \times 10^3$ g

    $$1.0 \times 10^3 \text{ g SiO}_2 \times \frac{1 \text{ mol}}{60.09 \text{ g}} = 16.64 \text{ mol SiO}_2$$

    From the balanced chemical equation, if 16.64 mol of $\text{SiO}_2$ were to react completely (an excess of carbon is present), then 16.64 mol of SiC should be produced (the coefficients of $\text{SiO}_2$ and SiC are the same).

    $$16.64 \text{ mol SiC} \times \frac{40.01 \text{ g}}{1 \text{ mol}} = 6.7 \times 10^2 \text{ g SiC} = 0.67 \text{ kg SiC}$$

60. If the reaction is performed in a solvent, the product may have a substantial solubility in the solvent; the reaction may come to equilibrium before the full yield of product is achieved (see Chapter 16); loss of product may occur through operator error.

62. $2\text{NaN}_3(s) \rightarrow 2\text{Na}(s) + 3\text{N}_2(g)$

    molar mass: $\text{NaN}_3$, 65.02 g; Na, 22.99 g

    $$10.5 \text{ g NaN}_3 \times \frac{1 \text{ mol NaN}_3}{65.02 \text{ g NaN}_3} = 0.161 \text{ mol NaN}_3$$

    $$0.161 \text{ mol NaN}_3 \times \frac{2 \text{ mol Na}}{2 \text{ mol NaN}_3} = 0.161 \text{ mol Na}$$

    $$0.161 \text{ mol Na} \times \frac{22.99 \text{ g Na}}{1 \text{ mol Na}} = 3.71 \text{ g Na theoretical yield}$$

    $$\% \text{ yield} = \frac{2.84 \text{ g actual yield}}{3.71 \text{ g theoretical yield}} \times 100 = 76.5\%$$

## Chapter 9: Chemical Quantities

64. $2LiOH(s) + CO_2(g) \rightarrow Li_2CO_3(s) + H_2O(g)$

    molar masses: LiOH, 23.95 g; $CO_2$, 44.01 g

    $$155 \text{ g LiOH} \times \frac{1 \text{ mol LiOH}}{23.95 \text{ g LiOH}} \times \frac{1 \text{ mol } CO_2}{2 \text{ mol LiOH}} \times \frac{44.01 \text{ g } CO_2}{1 \text{ mol } CO_2} = 142 \text{ g } CO_2$$

    As the cartridge has only absorbed 102 g $CO_2$ out of a total capacity of 142 g $CO_2$, the cartridge has absorbed

    $$\frac{102 \text{ g}}{142 \text{ g}} \times 100 = 71.8\% \text{ of its capacity.}$$

66. $2NH_3(g) + 3CuO(s) \rightarrow N_2(g) + 3Cu(s) + 3H_2O(g)$

    molar masses: $NH_3$, 17.034 g; CuO, 79.55 g; Cu, 63.55 g

    $$18.1 \text{ g } NH_3 \times \frac{1 \text{ mol } NH_3}{17.034 \text{ g } NH_3} = 1.06 \text{ mol } NH_3$$

    $$90.4 \text{ g } BaCl_2 \times \frac{1 \text{ mol CuO}}{79.55 \text{ g CuO}} = 1.14 \text{ mol CuO}$$

    CuO is the limiting reactant.

    $$1.14 \text{ mol CuO} \times \frac{3 \text{ mol Cu}}{3 \text{ mol CuO}} \times \frac{63.55 \text{ g Cu}}{1 \text{ mol Cu}} = 72.4 \text{ g Cu}$$

    $$\text{Percent yield} = \frac{\text{actual yield}}{\text{theoretical yield}} \times 100 = \frac{45.3 \text{ g}}{72.4 \text{ g}} \times 100 = 62.6\%$$

68. $NaCl(aq) + NH_3(aq) + H_2O(l) + CO_2(s) \rightarrow NH_4Cl(aq) + NaHCO_3(s)$

    molar masses: $NH_3$, 17.03 g; $CO_2$, 44.01 g; $NaHCO_3$, 84.01 g

    $$10.0 \text{ g } NH_3 \times \frac{1 \text{ mol}}{17.03 \text{ g}} = 0.5872 \text{ mol } NH_3$$

    $$15.0 \text{ g } CO_2 \times \frac{1 \text{ mol}}{44.01 \text{ g}} = 0.3408 \text{ mol } CO_2$$

    $CO_2$ is the limiting reactant.

    $$0.3408 \text{ mol } CO_2 \times \frac{1 \text{ mol } NaHCO_3}{1 \text{ mol } CO_2} = 0.3408 \text{ mol } NaHCO_3$$

    $$0.3408 \text{ mol } NaHCO_3 \times \frac{84.01 \text{ g}}{1 \text{ mol}} = 28.6 \text{ g } NaHCO_3$$

70. $C_6H_{12}O_6(s) + 6O_2(g) \rightarrow 6CO_2(g) + 6H_2O(g)$

    molar masses: glucose, 180.2 g; $CO_2$, 44.01 g

    $1.00 \text{ g glucose} \times \quad = 5.549 \times 10^{-3} \text{ mol glucose}$

Chapter 9: Chemical Quantities

$$5.549 \times 10^{-3} \text{ mol glucose} \times \frac{6 \text{ mol CO}_2}{1 \text{ mol glucose}} = 3.33 \times 10^{-2} \text{ mol CO}_2$$

$$3.33 \times 10^{-2} \text{ mol CO}_2 \times \frac{44.01 \text{ g}}{1 \text{ mol}} = 1.47 \text{ g CO}_2$$

72. $Ba^{2+}(aq) + SO_4^{2-}(aq) \rightarrow BaSO_4(s)$

    millimolar ionic masses: $Ba^{2+}$, 137.3 mg; $SO_4^{2-}$, 96.07 mg; $BaCl_2$, 208.2 mg

    $$150 \text{ mg SO}_4^{2-} \times \frac{1 \text{ mmol}}{96.07 \text{ mg}} = 1.56 \text{ millimol SO}_4^{2-}$$

    As barium ion and sulfate ion react on a 1:1 stoichiometric basis, then 1.56 millimol of barium ion is needed, which corresponds to 1.56 millimol of $BaCl_2$

    $$1.56 \text{ millimol BaCl}_2 \times \frac{208.2 \text{ mg}}{1 \text{ mmol}} = 325 \text{ milligrams BaCl}_2 \text{ needed}$$

74. a. $UO_2(s) + 4HF(aq) \rightarrow UF_4(aq) + 2H_2O(l)$

    One molecule (formula unit) of uranium(IV) oxide will combine with four molecules of hydrofluoric acid, producing one uranium(IV) fluoride molecule and two water molecules. One mole of uranium(IV) oxide will combine with four moles of hydrofluoric acid to produce one mole of uranium(IV) fluoride and two moles of water.

    b. $2NaC_2H_3O_2(aq) + H_2SO_4(aq) \rightarrow Na_2SO_4(aq) + 2HC_2H_3O_2(aq)$

    Two molecules (formula units) of sodium acetate react exactly with one molecule of sulfuric acid, producing one molecule (formula unit) of sodium sulfate and two molecules of acetic acid. Two moles of sodium acetate will combine with one mole of sulfuric acid, producing one mole of sodium sulfate and two moles of acetic acid.

    c. $Mg(s) + 2HCl(aq) \rightarrow MgCl_2(aq) + H_2(g)$

    One magnesium atom will react with two hydrochloric acid molecules (formula units) to produce one molecule (formula unit) of magnesium chloride and one molecule of hydrogen gas. One mole of magnesium will combine with two moles of hydrochloric acid, producing one mole of magnesium chloride and one mole of gaseous hydrogen.

    d. $B_2O_3(s) + 3H_2O(l) \rightarrow 2B(OH)_3(aq)$

    One molecule of diboron trioxide will react exactly with three molecules of water, producing two molecules of boron trihydroxide (boric acid). One mole of diboron trioxide will combine with three moles of water to produce two moles of boron trihydroxide (boric acid).

76. For $O_2$: $\left(\dfrac{5 \text{ mol O}_2}{1 \text{ mol C}_3H_8}\right)$   For $CO_2$: $\left(\dfrac{3 \text{ mol CO}_2}{1 \text{ mol C}_3H_8}\right)$   For $H_2O$: $\left(\dfrac{4 \text{ mol H}_2O}{1 \text{ mol C}_3H_8}\right)$

78. a. $NH_3(g) + HCl(g) \rightarrow NH_4Cl(s)$

    molar mass of $NH_3$ = 17.01 g

Chapter 9: Chemical Quantities

$$1.00 \text{ g NH}_3 \times \frac{1 \text{ mol}}{17.01 \text{ g}} = 0.0588 \text{ mol NH}_3$$

$$0.0588 \text{ mol NH}_3 \times \frac{1 \text{ mol NH}_4\text{Cl}}{1 \text{ mol NH}_3} = 0.0588 \text{ mol NH}_4\text{Cl}$$

b. $CaO(s) + CO_2(g) \rightarrow CaCO_3(s)$

molar mass CaO = 56.08 g

$$1.00 \text{ g CaO} \times \frac{1 \text{ mol}}{56.08 \text{ g}} = 0.0178 \text{ mol CaO}$$

$$0.0178 \text{ mol CaO} \times \frac{1 \text{ mol CaCO}_3}{1 \text{ mol CaO}} = 0.0178 \text{ mol CaCO}_3$$

c. $4Na(s) + O_2(g) \rightarrow 2Na_2O(s)$

molar mass Na = 22.99 g

$$1.00 \text{ g Na} \times \frac{1 \text{ mol}}{22.99 \text{ g}} = 0.0435 \text{ mol Na}$$

$$0.0435 \text{ mol Na} \times \frac{2 \text{ mol Na}_2\text{O}}{4 \text{ mol Na}} = 0.0217 \text{ mol Na}_2\text{O}$$

d. $2P(s) + 3Cl_2(g) \rightarrow 2PCl_3(l)$

molar mass P = 30.97 g

$$1.00 \text{ g P} \times \frac{1 \text{ mol}}{30.97 \text{ g}} = 0.0323 \text{ mol P}$$

$$0.0323 \text{ mol P} \times \frac{2 \text{ mol PCl}_3}{2 \text{ mol P}} = 0.0323 \text{ mol PCl}_3$$

80. a. molar mass $HNO_3$ = 63.0 g

$$5.0 \text{ mol HNO}_3 \times \frac{63.0 \text{ g}}{1 \text{ mol}} = 3.2 \times 10^2 \text{ g HNO}_3$$

b. molar mass Hg = 200.6 g

$$0.000305 \text{ mol Hg} \times \frac{200.6 \text{ g}}{1 \text{ mol}} = 0.0612 \text{ g Hg}$$

c. molar mass $K_2CrO_4$ = 194.2 g

$$2.31 \times 10^{-5} \text{ mol K}_2\text{CrO}_4 \times \frac{194.2 \text{ g}}{1 \text{ mol}} = 4.49 \times 10^{-3} \text{ g K}_2\text{CrO}_4$$

d. molar mass $AlCl_3$ = 133.3 g

$$10.5 \text{ mol AlCl}_3 \times \frac{133.3 \text{ g}}{1 \text{ mol}} = 1.40 \times 10^3 \text{ g AlCl}_3$$

e. molar mass $SF_6$ = 146.1 g

Chapter 9: Chemical Quantities

$$4.9 \times 10^4 \text{ mol SF}_6 \times \frac{146.1 \text{ g}}{1 \text{ mol}} = 7.2 \times 10^6 \text{ g SF}_6$$

f. molar mass $NH_3$ = 17.01 g

$$125 \text{ mol NH}_3 \times \frac{17.01 \text{ g}}{1 \text{ mol}} = 2.13 \times 10^3 \text{ g NH}_3$$

g. molar mass $Na_2O_2$ = 77.98 g

$$0.01205 \text{ mol Na}_2\text{O}_2 \times \frac{77.98 \text{ g}}{1 \text{ mol}} = 0.9397 \text{ g Na}_2\text{O}_2$$

82. $2SO_2(g) + O_2(g) \rightarrow 2SO_3(g)$

molar masses: $SO_2$, 64.07 g; $SO_3$, 80.07 g; 150 kg = $1.5 \times 10^5$ g

$$1.5 \times 10^5 \text{ g SO}_2 \times \frac{1 \text{ mol}}{64.07 \text{ g}} = 2.34 \times 10^3 \text{ mol SO}_2$$

$$2.34 \times 10^3 \text{ mol SO}_2 \times \frac{2 \text{ mol SO}_3}{2 \text{ mol SO}_2} = 2.34 \times 10^3 \text{ mol SO}_3$$

$$2.34 \times 10^3 \text{ mol SO}_3 \times \frac{80.07 \text{ g}}{1 \text{ mol}} = 1.9 \times 10^5 \text{ g SO}_3 = 1.9 \times 10^2 \text{ kg SO}_3$$

84. $2Na_2O_2(s) + 2H_2O(l) \rightarrow 4NaOH(aq) + O_2(g)$

molar masses: $Na_2O_2$, 77.98 g; $O_2$, 32.00 g

$$3.25 \text{ g Na}_2\text{O}_2 \times \frac{1 \text{ mol}}{77.98 \text{ g}} = 0.0417 \text{ mol Na}_2\text{O}_2$$

$$0.0417 \text{ mol Na}_2\text{O}_2 \times \frac{1 \text{ mol O}_2}{2 \text{ mol Na}_2\text{O}_2} = 0.0209 \text{ mol O}_2$$

$$0.0209 \text{ mol O}_2 \times \frac{32.00 \text{ g}}{1 \text{ mol}} = 0.667 \text{ g O}_2$$

86. $Zn(s) + 2HCl(aq) \rightarrow ZnCl_2(aq) + H_2(g)$

molar masses: Zn, 65.38 g; $H_2$, 2.016 g

$$2.50 \text{ g Zn} \times \frac{1 \text{ mol}}{65.38 \text{ g}} = 0.03824 \text{ mol Zn}$$

$$0.03824 \text{ mol Zn} \times \frac{1 \text{ mol H}_2}{1 \text{ mol Zn}} = 0.03824 \text{ mol H}_2$$

$$0.03824 \text{ mol H}_2 \times \frac{2.016 \text{ g}}{1 \text{ mol}} = 0.0771 \text{ g H}_2$$

88. a. $2Na(s) + Br_2(l) \rightarrow 2NaBr(s)$

molar masses: Na, 22.99 g; $Br_2$, 159.8 g; NaBr, 102.9 g

# Chapter 9: Chemical Quantities

$$5.0 \text{ g Na} \times \frac{1 \text{ mol}}{22.99 \text{ g}} = 0.2175 \text{ mol Na}$$

$$5.0 \text{ g Br}_2 \times \frac{1 \text{ mol}}{159.8 \text{ g}} = 0.03129 \text{ mol Br}_2$$

Intuitively, we would suspect that $Br_2$ is the limiting reactant, because there is much less $Br_2$ than Na on a mole basis. To *prove* that $Br_2$ is the limiting reactant, the following calculation is needed:

$$0.03129 \text{ mol Br}_2 \times \frac{2 \text{ mol Na}}{1 \text{ mol Br}_2} = 0.06258 \text{ mol Na}.$$

Clearly, there is more Na than this present, so $Br_2$ limits the reaction extent and the amount of NaBr formed.

$$0.03129 \text{ mol Br}_2 \times \frac{2 \text{ mol NaBr}}{1 \text{ mol Br}_2} = 0.06258 \text{ mol NaBr}$$

$$0.06258 \text{ mol NaBr} \times \frac{102.9 \text{ g}}{1 \text{ mol}} = 6.4 \text{ g NaBr}$$

b. $Zn(s) + CuSO_4(aq) \rightarrow ZnSO_4(aq) + Cu(s)$

molar masses: Zn, 65.38 g; Cu, 63.55 g; $ZnSO_4$, 161.5 g; $CuSO_4$, 159.6 g

$$5.0 \text{ g Zn} \times \frac{1 \text{ mol}}{65.38 \text{ g}} = 0.07648 \text{ mol Zn}$$

$$5.0 \text{ g CuSO}_4 \times \frac{1 \text{ mol}}{159.6 \text{ g}} = 0.03132 \text{ mol CuSO}_4$$

As the coefficients of Zn and $CuSO_4$ are the *same* in the balanced chemical equation, an equal number of moles of Zn and $CuSO_4$ would be needed for complete reaction. There is less $CuSO_4$ present, so $CuSO_4$ must be the limiting reactant.

$$0.03132 \text{ mol CuSO}_4 \times \frac{1 \text{ mol ZnSO}_4}{1 \text{ mol CuSO}_4} = 0.03132 \text{ mol ZnSO}_4$$

$$0.03132 \text{ mol ZnSO}_4 \times \frac{161.5 \text{ g}}{1 \text{ mol}} = 5.1 \text{ g ZnSO}_4$$

$$0.03132 \text{ mol CuSO}_4 \times \frac{1 \text{ mol Cu}}{1 \text{ mol CuSO}_4} = 0.03132 \text{ mol Cu}$$

$$0.03132 \text{ mol Cu} \times \frac{63.55 \text{ g}}{1 \text{ mol}} = 2.0 \text{ g Cu}$$

c. $NH_4Cl(aq) + NaOH(aq) \rightarrow NH_3(g) + H_2O(l) + NaCl(aq)$

molar masses: $NH_4Cl$, 53.49 g; NaOH, 40.00 g; $NH_3$, 17.03 g; $H_2O$, 18.02 g; NaCl, 58.44 g

$$5.0 \text{ g NH}_4\text{Cl} \times \frac{1 \text{ mol}}{53.49 \text{ g}} = 0.09348 \text{ mol NH}_4\text{Cl}$$

$$5.0 \text{ g NaOH} \times \frac{1 \text{ mol}}{40.00 \text{ g}} = 0.1250 \text{ mol NaOH}$$

As the coefficients of $NH_4Cl$ and NaOH are both *one* in the balanced chemical equation for the reaction, an equal number of moles of $NH_4Cl$ and NaOH would be needed for complete reaction. There is less $NH_4Cl$ present, so $NH_4Cl$ must be the limiting reactant.

As the coefficients of the products in the balanced chemical equation are also all *one*, if 0.09348 mol of $NH_4Cl$ (the limiting reactant) reacts completely, then 0.09348 mol of each product will be formed.

$$0.09348 \text{ mol NH}_3 \times \frac{17.03 \text{ g}}{1 \text{ mol}} = 1.6 \text{ g NH}_3$$

$$0.09348 \text{ mol H}_2\text{O} \times \frac{18.02 \text{ g}}{1 \text{ mol}} = 1.7 \text{ g H}_2\text{O}$$

$$0.09348 \text{ mol NaCl} \times \frac{58.44 \text{ g}}{1 \text{ mol}} = 5.5 \text{ g NaCl}$$

d. $Fe_2O_3(s) + 3CO(g) \rightarrow 2Fe(s) + 3CO_2(g)$

molar masses: $Fe_2O_3$, 159.7 g; CO, 28.01 g; Fe, 55.85 g; $CO_2$, 44.01 g

$$5.0 \text{ g Fe}_2\text{O}_3 \times \frac{1 \text{ mol}}{159.7 \text{ g}} = 0.03131 \text{ mol Fe}_2\text{O}_3$$

$$5.0 \text{ g CO} \times \frac{1 \text{ mol}}{28.01 \text{ g}} = 0.1785 \text{ mol CO}$$

Because there is considerably less $Fe_2O_3$ than CO on a mole basis, let's see if $Fe_2O_3$ is the limiting reactant.

$$0.03131 \text{ mol Fe}_2\text{O}_3 \times \frac{3 \text{ mol CO}}{1 \text{ mol Fe}_2\text{O}_3} = 0.09393 \text{ mol CO}$$

As there is 0.1785 mol of CO present, but we have determined that only 0.09393 mol CO would be needed to react with all the $Fe_2O_3$ present, then $Fe_2O_3$ must be the limiting reactant. CO is present in excess.

$$0.03131 \text{ mol Fe}_2\text{O}_3 \times \frac{2 \text{ mol Fe}}{1 \text{ mol Fe}_2\text{O}_3} \times \frac{55.85 \text{ g Fe}}{1 \text{ mol Fe}} = 3.5 \text{ g Fe}$$

$$0.03131 \text{ mol Fe}_2\text{O}_3 \times \frac{3 \text{ mol CO}_2}{1 \text{ mol Fe}_2\text{O}_3} \times \frac{44.01 \text{ g CO}_2}{1 \text{ mol CO}_2} = 4.1 \text{ g CO}_2$$

90. $N_2H_4(l) + O_2(g) \rightarrow N_2(g) + 2H_2O(g)$

molar masses: $N_2H_4$, 32.05 g; $O_2$, 32.00 g; $N_2$, 28.02 g; $H_2O$, 18.02 g

$$20.0 \text{ g N}_2\text{H}_4 \times \frac{1 \text{ mol}}{32.05 \text{ g}} = 0.624 \text{ mol N}_2\text{H}_4$$

$$20.0 \text{ g O}_2 \times \frac{1 \text{ mol}}{32.00 \text{ g}} = 0.625 \text{ mol O}_2$$

Chapter 9: Chemical Quantities

The two reactants are present in nearly the required ratio for complete reaction (due to the 1:1 stoichiometry of the reaction and the very similar molar masses of the substances). We will consider $N_2H_4$ as the limiting reactant in the following calculations.

$$0.624 \text{ mol } N_2H_4 \times \frac{1 \text{ mol } N_2}{1 \text{ mol } N_2H_4} \times \frac{28.02 \text{ g } N_2}{1 \text{ mol } N_2} = 17.5 \text{ g } N_2$$

$$0.624 \text{ mol } N_2H_4 \times \frac{2 \text{ mol } H_2O}{1 \text{ mol } N_2H_4} \times \frac{18.02 \text{ g } H_2O}{1 \text{ mol } H_2O} = 22.5 \text{ g } H_2O$$

92. $12.5 \text{ g theory} \times \dfrac{40 \text{ g actual}}{100 \text{ g theory}} = 5.0 \text{ g}$

94. The balanced equation is: $2NaNO_3 \rightarrow 2NaNO_2 + O_2$

$$0.2339 \text{ g } NaNO_2 \times \frac{1 \text{ mol } NaNO_2}{69.00 \text{ g } NaNO_2} = 0.003390 \text{ mol } NaNO_2$$

$$0.003390 \text{ mol } NaNO_2 \times \frac{2 \text{ mol } NaNO_3}{2 \text{ mol } NaNO_2} = 0.003390 \text{ mol } NaNO_3$$

$$0.003390 \text{ mol } NaNO_3 \times \frac{85.00 \text{ g } NaNO_3}{1 \text{ mol } NaNO_3} = 0.28815 \text{ g } NaNO_3$$

$$\% \text{ } NaNO_3 = \frac{0.28815 \text{ g}}{0.4230 \text{ g}} \times 100 = 68.12\%$$

96. The balanced equation is: $Fe_2O_3(s) + 2Al(s) \rightarrow 2Fe(l) + Al_2O_3(s)$

   a. $25.69 \text{ g Fe} \times \dfrac{1 \text{ mol Fe}}{55.85 \text{ g Fe}} = 0.4600 \text{ mol Fe}$

   $0.4600 \text{ mol Fe} \times \dfrac{1 \text{ mol } Fe_2O_3}{2 \text{ mol Fe}} = 0.2300 \text{ mol } Fe_2O_3$

   $0.2300 \text{ mol } Fe_2O_3 \times \dfrac{159.7 \text{ g } Fe_2O_3}{1 \text{ mol } Fe_2O_3} = 36.73 \text{ g } Fe_2O_3$

   b. $0.4600 \text{ mol Fe} \times \dfrac{2 \text{ mol Al}}{2 \text{ mol Fe}} = 0.4600 \text{ mol Al}$

   $0.4600 \text{ mol Al} \times \dfrac{26.98 \text{ g Al}}{1 \text{ mol Al}} = 12.41 \text{ g Al}$

   c. $0.4600 \text{ mol Fe} \times \dfrac{1 \text{ mol } Al_2O_3}{2 \text{ mol Fe}} = 0.2300 \text{ mol } Al_2O_3$

   $0.2300 \text{ mol } Al_2O_3 \times \dfrac{101.96 \text{ g } Al_2O_3}{1 \text{ mol } Al_2O_3} = 23.45 \text{ g } Al_2O_3$

## Chapter 9: Chemical Quantities

98. The balanced equation is: $2NH_3(g) + 2Na(s) \rightarrow 2NaNH_2(s) + H_2(g)$

$$32.8 \text{ g NH}_3 \times \frac{1 \text{ mol NH}_3}{17.034 \text{ g NH}_3} = 1.93 \text{ mol NH}_3$$

$$16.6 \text{ g Na} \times \frac{1 \text{ mol Na}}{22.99 \text{ g Na}} = 0.722 \text{ mol Na}$$

Na is the limiting reactant (only 0.722 mol $NH_3$ needed).

$$0.722 \text{ mol Na} \times \frac{2 \text{ mol NaNH}_2}{2 \text{ mol Na}} \times \frac{39.016 \text{ g NaNH}_2}{1 \text{ mol NaNH}_2} = 28.2 \text{ g NaNH}_2$$

$$0.722 \text{ mol Na} \times \frac{1 \text{ mol H}_2}{2 \text{ mol Na}} \times \frac{2.016 \text{ g H}_2}{1 \text{ mol H}_2} = 0.728 \text{ g H}_2$$

100. $2C_3H_6(g) + 2NH_3(g) + 3O_2(g) \rightarrow 2C_3H_3N(g) + 6H_2O(g)$

   a. $$5.23 \times 10^2 \text{ g C}_3\text{H}_6 \times \frac{1 \text{ mol C}_3\text{H}_6}{42.078 \text{ g C}_3\text{H}_6} = 12.4 \text{ mol C}_3\text{H}_6$$

   $$12.4 \text{ mol C}_3\text{H}_6 \times \frac{2 \text{ mol C}_3\text{H}_3\text{N}}{2 \text{ mol C}_3\text{H}_6} \times \frac{53.064 \text{ g C}_3\text{H}_3\text{N}}{1 \text{ mol C}_3\text{H}_3\text{N}} = 658 \text{ g C}_3\text{H}_3\text{N}$$

   $$5.00 \times 10^2 \text{ g NH}_3 \times \frac{1 \text{ mol NH}_3}{17.034 \text{ g NH}_3} = 29.4 \text{ mol NH}_3$$

   $$29.4 \text{ mol NH}_3 \times \frac{2 \text{ mol C}_3\text{H}_3\text{N}}{2 \text{ mol NH}_3} \times \frac{53.064 \text{ g C}_3\text{H}_3\text{N}}{1 \text{ mol C}_3\text{H}_3\text{N}} = 1560 \text{ g C}_3\text{H}_3\text{N}$$

   $$1.00 \times 10^3 \text{ g O}_2 \times \frac{1 \text{ mol O}_2}{32.00 \text{ g O}_2} = 31.25 \text{ mol O}_2$$

   $$31.25 \text{ mol O}_2 \times \frac{2 \text{ mol C}_3\text{H}_3\text{N}}{3 \text{ mol O}_2} \times \frac{53.064 \text{ g C}_3\text{H}_3\text{N}}{1 \text{ mol C}_3\text{H}_3\text{N}} = 1110 \text{ g C}_3\text{H}_3\text{N}$$

   Thus, $C_3H_6$ is the limiting reactant and 658 g $C_3H_3N$ is produced (660. g $C_3H_3N$ if decimals are carried through).

   b. $$12.4 \text{ mol C}_3\text{H}_6 \times \frac{6 \text{ mol H}_2\text{O}}{2 \text{ mol C}_3\text{H}_6} \times \frac{18.016 \text{ g H}_2\text{O}}{1 \text{ mol H}_2\text{O}} = 670. \text{ g H}_2\text{O} \text{ (672 g H}_2\text{O if decimals are carried through)}$$

   c. $C_3H_6$: 0 g (limiting reactant and is used up)

   $NH_3$:

   $$12.4 \text{ mol C}_3\text{H}_6 \times \frac{2 \text{ mol NH}_3}{2 \text{ mol C}_3\text{H}_6} = 12.4 \text{ mol NH}_3 \text{ used up}$$

   29.4 mol $NH_3$ – 12.4 mol $NH_3$ = 17.0 mol $NH_3$ leftover

$17.0 \text{ mol NH}_3 \times \dfrac{17.034 \text{ g NH}_3}{1 \text{ mol NH}_3} = 290. \text{ g NH}_3$ (289 g NH$_3$ if decimals are carried through)

O$_2$:

$12.4 \text{ mol C}_3\text{H}_6 \times \dfrac{3 \text{ mol O}_2}{2 \text{ mol C}_3\text{H}_6} = 18.6 \text{ mol O}_2$ used up

$31.25 \text{ mol O}_2 - 18.6 \text{ mol O}_2 = 12.65 \text{ mol O}_2$ leftover

$12.65 \text{ mol O}_2 \times \dfrac{32.00 \text{ g O}_2}{1 \text{ mol O}_2} = 405 \text{ g O}_2$

# CUMULATIVE REVIEW

# Chapters 8 and 9

2. On a microscopic basis, one mole of a substance represents Avogadro's number ($6.022 \times 10^{23}$) of individual units (atoms or molecules) of the substance. On a macroscopic, more practical basis, one mole of a substance represents the amount of substance present when the molar mass of the substance in grams is taken (for example 12.01 g of carbon will be one mole of carbon). Chemists have chosen these definitions so that there will be a simple relationship between measurable amounts of substances (grams) and the actual number of atoms or molecules present, and so that the number of particles present in samples of *different* substances can easily be compared. For example, it is known that carbon and oxygen react by the reaction

$$C(s) + O_2(g) \rightarrow CO_2(g).$$

Chemists understand this equation to mean that one carbon atom reacts with one oxygen molecule to produce one molecule of carbon dioxide, and also that one mole (12.01 g) of carbon will react with one mole (32.00 g) of oxygen to produce one mole (44.01 g) of carbon dioxide.

4. The molar mass of a compound is the mass in grams of one mole of the compound ($6.022 \times 10^{23}$ molecules of the compound), and is calculated by summing the average atomic masses of all the atoms present in a molecule of the compound. For example, a molecule of the compound $H_3PO_4$ contains three hydrogen atoms, one phosphorus atom, and four oxygen atoms: the molar mass is obtained by adding up the average atomic masses of these atoms: molar mass $H_3PO_4$ = 3(1.008 g) + 1(30.97 g) + 4(16.00 g) = 97.99 g

6. The empirical formula of a compound represents the lowest ratio of the relative number of atoms of each type present in a molecule of the compound, whereas the molecular formula represents the actual number of atoms of each type present in a real molecule of the compound. For example, both acetylene (molecular formula $C_2H_2$) and benzene (molecular formula $C_6H_6$) have the same relative number of carbon and hydrogen atoms (one hydrogen for each carbon atom), and so have the same empirical formula (CH). Once the empirical formula of a compound has been determined, it is also necessary to determine the molar mass of the compound before the actual molecular formula can be calculated. As real molecules cannot contain fractional parts of atoms, the molecular formula is always a whole number multiple of the empirical formula. For the examples above, the molecular formula of acetylene is twice the empirical formula, and the molecular formula of benzene is six times the empirical formula (both factors are integers).

8. In question 7, suppose we chose to calculate the percentage composition of phosphoric acid, $H_3PO_4$: 3.086% H, 31.60% P, and 65.31% O. We could convert this percentage composition data into "experimental" data by first choosing a mass of sample to be "analyzed", and then calculating what mass of each element is present in this size sample using the percentage of each element. For example, suppose we choose our sample to have a mass of 2.417 g. Then the masses of H, P, and O present in this sample would be given by the following:

$$\text{g H} = (2.417 \text{ g sample}) \times \frac{3.086 \text{ g H}}{100.0 \text{ g sample}} = 0.07459 \text{ g H}$$

Review:   Chapters 8 and 9

$$\text{g P} = (2.417 \text{ g sample}) \times \frac{31.60 \text{ g P}}{100.0 \text{ g sample}} = 0.7638 \text{ g P}$$

$$\text{g O} = (2.417 \text{ g sample}) \times \frac{65.31 \text{ g O}}{100.0 \text{ g sample}} = 1.579 \text{ g O}$$

Note that (0.07459 g + 0.7638 g + 1.579 g) = 2.41739 = 2.417 g.

So our new problem could be worded as follows: "A 2.417 g sample of a compound has been analyzed and was found to contain 0.07459 g H, 0.7638 g of P, and 1.579 g of oxygen. Calculate the empirical formula of the compound".

$$\text{mol H} = (0.07459 \text{ g H}) \times \frac{1 \text{ mol H}}{1.008 \text{ g H}} = 0.07400 \text{ mol H}$$

$$\text{mol P} = (0.7638 \text{ g P}) \times \frac{1 \text{ mol P}}{30.97 \text{ g P}} = 0.02466 \text{ mol P}$$

$$\text{mol O} = (1.579 \text{ g O}) \times \frac{1 \text{ mol O}}{16.00 \text{ g O}} = 0.09869 \text{ mol O}$$

Dividing each of these numbers of moles by the smallest number of moles (0.02466 mol P) gives the following:

$$\frac{0.07400 \text{ mol H}}{0.02466} = 3.001 \text{ mol H}$$

$$\frac{0.02466 \text{ mol P}}{0.02466} = 1.000 \text{ mol P}$$

$$\frac{0.09869 \text{ mol O}}{0.02466} = 4.002 \text{ mol O}$$

The empirical formula is (not surprisingly) just $H_3PO_4$!

10. The mole ratios for a reaction are based on the *coefficients* of the balanced chemical equation for the reaction: these coefficients show in what proportions molecules (or moles of molecules) combine. For a given amount of propane, the following mole ratios could be constructed, which would enable you to calculate the number of moles of each product, or of the second reactant, that would be involved.

$$C_3H_8(g) + 5O_2(g) \rightarrow 3CO_2(g) + 4H_2O(g)$$

for $O_2$: $\dfrac{5 \text{ mol } O_2}{1 \text{ mol } C_3H_8}$; $0.55 \text{ mol } C_3H_8 \times \dfrac{5 \text{ mol } O_2}{1 \text{ mol } C_3H_8} = 2.8 \ (2.75) \text{ mol } O_2$

for $CO_2$: $\dfrac{3 \text{ mol } CO_2}{1 \text{ mol } C_3H_8}$; $0.55 \text{ mol } C_3H_8 \times \dfrac{3 \text{ mol } CO_2}{1 \text{ mol } C_3H_8} = 1.7 \ (1.65) \text{ mol } CO_2$

for $H_2O$: $\dfrac{4 \text{ mol } H_2O}{1 \text{ mol } C_3H_8}$; $0.55 \text{ mol } C_3H_8 \times \dfrac{4 \text{ mol } H_2O}{1 \text{ mol } C_3H_8} = 2.2 \text{ mol } H_2O$

Review: Chapters 8 and 9

12. Although we can calculate specifically the exact amounts of each reactant needed for a chemical reaction, oftentimes reaction mixtures are prepared using more or less arbitrary amounts of the reagents. However, regardless of how much of each reagent may be used for a reaction, the substances still react stoichiometrically, according to the mole ratios derived from the balanced chemical equation for the reaction. When arbitrary amounts of reactants are used, there will be one reactant which, stoichiometrically, is present in the least amount. This substance is called the *limiting reactant* for the experiment. It is the limiting reactant that controls how much product is formed, regardless of how much of the other reactants are present. The limiting reactant limits the amount of product that can form in the experiment, because once the limiting reactant has reacted completely, the reaction must stop. We say that the other reactants in the experiment are present in excess, which means that a portion of these reactants will still be present unchanged after the reaction has ended and the limiting reactant has been used up completely.

14. The *theoretical yield* for an experiment is the mass of product calculated based on the limiting reactant for the experiment being completely consumed. The *actual yield* for an experiment is the mass of product actually collected by the experimenter. Obviously, any experiment is restricted by the skills of the experimenter and by the inherent limitations of the experimental method being used. For these reasons, the actual yield is often *less* than the theoretical yield (most scientific writers report the actual or percentage yield for their experiments as an indication of the usefulness of their experiments). Although one would expect that the actual yield should never be more than the theoretical yield, in real experiments, sometimes this happens: however, an actual yield greater than a theoretical yield is usually taken to mean that something is *wrong* in either the experiment (for example, impurities may be present, or the reaction may not occur as envisioned) or in the calculations.

16. % element X = $\dfrac{\text{mass of element X in compound}}{\text{molar mass of compound}} \times 100$

   a. 92.26% C      b. 32.37% Na
   c. 15.77% C      d. 20.24% Al
   e. 88.82% Cu     f. 79.89% Cu
   g. 71.06% Co     h. 40.00% C

18. a. molar masses: SiC, 40.10 g; SiCl$_4$, 169.9 g

   $12.5 \text{ g SiC} \times \dfrac{1 \text{ mol}}{40.10 \text{ g}} = 0.3117 \text{ mol SiC}$

   for SiCl$_4$: $0.3117 \text{ mol SiC} \times \dfrac{1 \text{ mol SiCl}_4}{1 \text{ mol SiC}} \times \dfrac{169.9 \text{ g SiCl}_4}{1 \text{ mol SiCl}_4} = 53.0 \text{ g SiCl}_4$

   for C: $0.3117 \text{ mol SiC} \times \dfrac{1 \text{ mol C}}{1 \text{ mol SiC}} \times \dfrac{12.01 \text{ g C}}{1 \text{ mol C}} = 3.75 \text{ g C}$

   b. molar masses: Li$_2$O, 29.88 g; LiOH, 23.95 g

   $12.5 \text{ g Li}_2\text{O} \times \dfrac{1 \text{ mol}}{29.88 \text{ g}} = 0.4183 \text{ mol Li}_2\text{O}$

Review: Chapters 8 and 9

$$0.4183 \text{ mol Li}_2\text{O} \times \frac{2 \text{ mol LiOH}}{1 \text{ mol Li}_2\text{O}} \times \frac{23.95 \text{ g LiOH}}{1 \text{ mol LiOH}} = 20.0 \text{ g LiOH}$$

c. molar masses: $Na_2O_2$, 77.98 g; NaOH, 40.00 g; $O_2$, 32.00 g

$$12.5 \text{ g} \times \frac{1 \text{ mol}}{77.98 \text{ g}} = 0.1603 \text{ mol Na}_2\text{O}_2$$

for NaOH: $0.1603 \text{ mol Na}_2\text{O}_2 \times \dfrac{4 \text{ mol NaOH}}{2 \text{ mol Na}_2\text{O}_2} \times \dfrac{40.00 \text{ g NaOH}}{1 \text{ mol NaOH}} = 12.8 \text{ g NaOH}$

for $O_2$: $0.1603 \text{ mol Na}_2\text{O}_2 \times \dfrac{1 \text{ mol O}_2}{2 \text{ mol Na}_2\text{O}_2} \times \dfrac{32.00 \text{ g O}_2}{1 \text{ mol O}_2} = 2.56 \text{ g O}_2$

d. molar masses: $SnO_2$, 150.7 g; Sn, 118.7 g; $H_2O$, 18.02 g

$$12.5 \text{ g SnO}_2 \times \frac{1 \text{ mol}}{150.7 \text{ g}} = 0.08295 \text{ mol SnO}_2$$

for Sn: $0.08295 \text{ mol SnO}_2 \times \dfrac{1 \text{ mol Sn}}{1 \text{ mol SnO}_2} \times \dfrac{118.7 \text{ g Sn}}{1 \text{ mol Sn}} = 9.84 \text{ g Sn}$

for $H_2O$: $0.08295 \text{ mol SnO}_2 \times \dfrac{2 \text{ mol H}_2\text{O}}{1 \text{ mol SnO}_2} \times \dfrac{18.02 \text{ g H}_2\text{O}}{1 \text{ mol H}_2\text{O}} = 2.99 \text{ g H}_2\text{O}$

20. a. $CaC_2(s) + 2H_2O(l) \rightarrow C_2H_2(g) + Ca(OH)_2(aq)$

    b. $H_2O$ is the limiting reactant.

    c. mass before the reaction = mass after the reaction

    mass before the reaction = 100.0 g + 50.0 g = 150.0 g

    102.8 g $Ca(OH)_2$ produced; 36.13 g $C_2H_2$ produced; 11.06 g $CaC_2$ leftover

    mass after the reaction = 102.8 g + 36.13 g + 11.06 g = 150.0 g

# CHAPTER 10

# Energy

2. Potential energy is energy due to position or composition. A stone at the top of a hill possesses potential energy since the stone may eventually roll down the hill. A gallon of gasoline possesses potential energy since heat will be released when the gasoline is burned.

4. The total energy of the universe is *constant*.

6. Ball A initially possesses potential energy by virtue of its position at the top of the hill. As Ball A rolls down the hill, its potential energy is converted to kinetic energy and frictional (heat) energy. When Ball A reaches the bottom of the hill and hits Ball B, it transfers its kinetic energy to Ball B. Ball A then has only the potential energy corresponding to its new position.

8. The hot tea is at a higher temperature, which means the particles in the hot tea have higher average kinetic energies. When the tea spills on the skin, energy flows from the hot tea to the skin, until the tea and skin are at the same temperature. This sudden inflow of energy causes the burn.

10. Temperature is the concept by which we express the thermal energy contained in a sample. We cannot measure the motions of the particles/kinetic energy in a sample of matter directly. We know, however, that if two objects are at different temperatures, the one with the higher temperature has molecules that have higher average kinetic energies than the object at the lower temperature.

12. When the chemical system evolves energy, the energy evolved from the reacting chemicals is transferred to the surroundings.

14. a. Endothermic; Energy is required to dissolve the solid KBr in water.

    b. Exothermic; Energy is released in combustions reactions.

    c. Exothermic; Energy is released when concentrated sulfuric acid dissociates in water.

    d. Endothermic; Energy is required to change the water from a liquid to a gas.

16. internal

18. gaining

20. $\Delta E = q + w = +215 \text{ kJ} + 116 \text{ kJ} = 331 \text{ kJ}$

22. a. $\dfrac{1 \text{ J}}{4.184 \text{ cal}}$

    b. $\dfrac{4.184 \text{ cal}}{1 \text{ J}}$

Chapter 10:     Energy

      c.    $\dfrac{1 \text{ kcal}}{1000 \text{ cal}}$

      d.    $\dfrac{1000 \text{ J}}{1 \text{ kJ}}$

24.    6540 J = 6.54 kJ for 10 times more water

26.    a.    $8254 \text{ cal} \times \dfrac{1 \text{ kcal}}{1000 \text{ cal}} = 8.254 \text{ kcal}$

      b.    $41.5 \text{ cal} \times \dfrac{1 \text{ kcal}}{1000 \text{ cal}} = 0.0415 \text{ kcal}$

      c.    $8.231 \times 10^3 \times \dfrac{1 \text{ kcal}}{1000 \text{ cal}} = 8.231 \text{ kcal}$

      d.    $752{,}900 \text{ cal} \times \dfrac{1 \text{ kcal}}{1000 \text{ cal}} = 752.9 \text{ kcal}$

28.    a.    $7845 \text{ cal} \times \dfrac{4.184 \text{ J}}{1 \text{ cal}} = 32820 \text{ J}$

          $32820 \text{ J} \times \dfrac{1 \text{ kJ}}{1000 \text{ J}} = 32.82 \text{ kJ}$

      b.    $4.55 \times 10^4 \text{ cal} \times \dfrac{4.184 \text{ J}}{1 \text{ cal}} = 1.90 \times 10^5 \text{ J}$

          $1.90 \times 10^5 \text{ J} \times \dfrac{1 \text{ kJ}}{1000 \text{ J}} = 190. \text{ kJ}$

      c.    $62.142 \text{ kcal} \times \dfrac{4.184 \text{ kJ}}{1 \text{ kcal}} = 260.0 \text{ kJ}$

          $260.0 \text{ kJ} \times \dfrac{1000 \text{ J}}{1 \text{ kJ}} = 2.600 \times 10^5 \text{ J}$

      d.    $43024 \text{ cal} \times \dfrac{4.184 \text{ J}}{1 \text{ cal}} = 1.800 \times 10^5 \text{ J}$

          $1.800 \times 10^5 \text{ J} \times \dfrac{1 \text{ kJ}}{1000 \text{ J}} = 180.0 \text{ kJ}$

30.    a.    $91.74 \text{ kcal} \times \dfrac{1000 \text{ cal}}{1 \text{ kcal}} = 9.174 \times 10^4 \text{ cal}$

      b.    $1.781 \text{ kJ} \times \dfrac{1000 \text{ J}}{1 \text{ kJ}} \times \dfrac{1 \text{ cal}}{4.184 \text{ J}} = 425.7 \text{ cal}$

      c.    $4.318 \times 10^3 \text{ J} \times \dfrac{1 \text{ cal}}{4.184 \text{ J}} \times \dfrac{1 \text{ kcal}}{1000 \text{ cal}} = 1.032 \text{ kcal}$

Chapter 10:  Energy

d. $9.173 \times 10^4 \text{ cal} \times \dfrac{4.184 \text{ J}}{1 \text{ cal}} \times \dfrac{1 \text{ kJ}}{1000 \text{ J}} = 383.8 \text{ kJ}$

32. $Q = s \times m \times \Delta T$

Specific heat capacity of aluminum is 0.89 J/g°C from Table 10.1.

$Q = (0.89 \text{ J/g°C}) \times (42.7 \text{ g}) \times (15.2°C) = 5.8 \times 10^2$ J (only two significant figures are justified).

34. $Q = s \times m \times \Delta T$

Specific heat capacity of mercury is 0.14 J/g°C

100. J = (0.14 J/g°C) × 25 g × $\Delta T$

$\Delta T$ = 28.6°C = 29°C

36. The reaction is exothermic: the chemical reaction liberates heat energy and warms up the beverage.

38. $Q = s \times m \times \Delta T$

specific heat capacity of water = 4.184 J/g°C

$Q = 4.184$ J/g°C × 100.0 g × 35°C = $1.46 \times 10^4$ J (~15 kJ to two significant figures)

40. calorimeter

42. a. molar mass of S = 32.07 g

$1.00 \text{ g S} \times \dfrac{1 \text{ mol S}}{32.07 \text{ g S}} = 0.0312 \text{ mol S}$

From the balanced chemical equation, combustion of 0.0312 mol S would produce 0.0312 mol SO$_2$

$0.0312 \text{ mol} \times \dfrac{-296 \text{ kJ}}{\text{mol}} = -9.23 \text{ kJ}$

b. From the balanced chemical equation, combustion of 0.0312 mol S would produce 0.0312 mol SO$_2$

$0.501 \text{ mol} \times \dfrac{-296 \text{ kJ}}{\text{mol}} = -148 \text{ kJ}$

c. The enthalpy change would be the same in magnitude, but opposite in sign = +296 kJ/mol

44. a. molar mass of ethanol = 46.07 g; we will assume that 1360 has only 3 significant figures.

$-\dfrac{1360 \text{ kJ}}{1 \text{ mol}} \times \dfrac{1 \text{ mol}}{46.07 \text{ g}} = -29.5 \text{ kJ/g}$

b. Since energy is released by the combustion, $\Delta H$ is negative: $\Delta H = -1360$ kJ

Chapter 10:    Energy

      c.    In the reaction as written, three moles of water vapor are produced when one mole of ethanol reacts.

$$\frac{1360 \text{ kJ}}{1 \text{ mol } C_2H_5OH} \times \frac{1 \text{ mol } C_2H_5OH}{3 \text{ mol } H_2O} = 453 \text{ kJ/mol } H_2O$$

46. The desired equation $2C(s) + O_2(g) \rightarrow 2CO(g)$ can be generated by taking twice the first equation and adding it to the reverse of the second equation:

    $2 \times [C(s) + O_2(g) \rightarrow CO_2(g)]$      $\Delta H = 2 \times -393 \text{ kJ} = -786 \text{ kJ}$ (equation doubled)

    $2CO_2(g) \rightarrow 2CO(g) + O_2(g)$      $\Delta H = -(-566 \text{ kJ}) = +566 \text{ kJ}$ (equation reversed)

    $2C(s) + O_2(g) \rightarrow 2CO(g)$      $\Delta H = (-786) + (+566) = -220 \text{ kJ}$

48. The desired equation can be generated as follows:

    $2CO_2(g) + H_2O(l) \rightarrow C_2H_2(g) + 5/2 O_2(g)$      $\Delta H = -(-1300. \text{ kJ}) = 1300. \text{ kJ}$

    $2 \times [C(s) + O_2(g) \rightarrow CO_2(g)]$      $\Delta H = 2 \times -394 \text{ kJ} = -788 \text{ kJ}$

    $H_2(g) + 1/2 O_2(g) \rightarrow H_2O(l)$      $\Delta H = -286 \text{ kJ}$

    $2C(s) + H_2(g) \rightarrow C_2H_2(g)$      $\Delta H = (1300.) + (-788 \text{ kJ}) + (-286 \text{ kJ}) = 226 \text{ kJ}$

50. Once everything in the universe is at the same temperature, no further thermodynamic work can be done. Even though the total energy of the universe will be the same, the energy will have been dispersed evenly making it effectively useless.

52. Concentrated sources of energy, such as petroleum, are being used so as to disperse the energy they contain, making it unavailable for further use.

54. Petroleum consists mainly of hydrocarbons, which are molecules containing chains of carbon atoms with hydrogen atoms attached to the chains. The fractions are based on the number of carbon atoms in the chains: for example, gasoline is a mixture of hydrocarbons with 5–10 carbon atoms in the chains, whereas asphalt is a mixture of hydrocarbons with 25 or more carbon atoms in the chains. Different fractions have different physical properties and uses, but all can be combusted to produce energy. See Table 10.3

56. Tetraethyl lead was used as an additive for gasoline to promote smoother running of engines. It is no longer widely used because of concerns about the lead being released to the environment as the leaded gasoline is burned.

58. The greenhouse effect is a warming effect due to the presence of gases in the atmosphere which absorb infrared radiation that has reached the earth from the sun, and do not allow it to pass back into space. A limited greenhouse effect is desirable because it moderates the temperature changes in the atmosphere that would otherwise be more drastic between daytime when the sun is shining and nighttime. Having too high a concentration of greenhouse gases, however, will elevate the temperature of the earth too much, affecting climate, crops, the polar ice caps, temperature of the oceans, and so on. Carbon dioxide produced by combustion reactions is our greatest concern as a greenhouse gas.

60. If a proposed reaction involves either or both of those phenomena, the reaction will tend to be favorable.

Chapter 10:  Energy

62. Formation of a solid precipitate represents a concentration of matter.

64. The molecules in liquid water are moving around freely, and are therefore more "disordered" than when the molecules are held rigidly in a solid lattice in ice. The entropy increases during melting.

66. Statements (a), (b), (c), and (d) are all true.

68. Temperature increase = 75.0 – 22.3 = 52.7°C

$$145 \text{ g} \times 4.184 \frac{\text{J}}{\text{g °C}} \times 52.7°C \times \frac{1 \text{ cal}}{4.184 \text{ J}} = 7641.5 \text{ cal} = 7.65 \text{ kcal}$$

70. From Table 10.1, the specific heat capacity of iron is 0.45 J/g°C.

$$Q = s \times m \times \Delta T$$

$$Q = 0.45 \frac{\text{J}}{\text{g°C}} \times 25.1 \text{ g} \times 17.5°C = 197.7 \text{ J} = 2.0 \times 10^2 \text{ J (two significant figures)}$$

72. 2.5 kg water = 2500 g

Temperature change = 55.0 – 18.5 = 36.5°C

$$Q = s \times m \times \Delta T$$

$$Q = 4.184 \text{ J/g°C} \times 2500 \text{ g} \times 36.5°C = 3.8 \times 10^5 \text{ J}$$

74. Let $T_f$ represent the final temperature reached by the system.

For the hot water, heat lost = 50.0 g × 4.184 J/g°C × (100. – $T_f$°C)

For the cold water, heat gained = 50.0 g × 4.184 J/g°C × ($T_f$ – 25°C)

The heat lost by the hot water must *equal* the heat gained by the cold water; therefore

50.0 g × 4.184 J/g°C × (100. – $T_f$°C) = 50.0 g × 4.184 J/g°C × ($T_f$ – 25°C)

Solving this equation for $T_f$ gives $T_f$ = 62.5°C = 63°C

76. $Q = m \times s \times \Delta T$

$$Q = 7.24 \text{ kJ} \times \frac{1000 \text{ J}}{1 \text{ kJ}} = 7240 \text{ J}$$

$$s = \frac{Q}{m \times \Delta T} = \frac{7240 \text{ J}}{(952 \text{ g})(10.7°C)} = 0.711 \text{ J/g°C}$$

78. $q = –213$ kJ (heat is released) and $\Delta E = –45$ kJ

$\Delta E = q + w$

–45 kJ = –213 kJ + w

w = –45 kJ + 213 kJ = +168 kJ

Chapter 10: Energy

80. $5.00 \text{ g C}_3\text{H}_8 \times \dfrac{1 \text{ mol C}_3\text{H}_8}{44.09 \text{ g C}_3\text{H}_8} \times \dfrac{-2221 \text{ kJ}}{\text{mol}} = -252 \text{ kJ}$

82. Reversing the first equation and dividing by 6 we get

$\tfrac{3}{6}\text{D} \rightarrow \tfrac{3}{6}\text{A} + \text{B}$ $\quad\quad \Delta H = +403 \text{ kJ}/6$

or $\tfrac{1}{2}\text{D} \rightarrow \tfrac{1}{2}\text{A} + \text{B}$ $\quad\quad \Delta H = +67.2 \text{ kJ}$

Dividing the second equation by 2 we get

$\tfrac{1}{2}\text{E} + \text{F} \rightarrow \tfrac{1}{2}\text{A}$ $\quad\quad \Delta H = -105.2 \text{ kJ}/2 = -52.6 \text{ kJ}$

Dividing the third equation by 2 we get

$\tfrac{1}{2}\text{C} \rightarrow \tfrac{1}{2}\text{E} + \tfrac{3}{2}\text{D}$ $\quad\quad \Delta H = +64.8 \text{ kJ}/2 = +32.4 \text{ kJ}$

Adding these equations together we get

$\tfrac{1}{2}\text{C} + \text{F} \rightarrow \text{A} + \text{B} + \text{D}$ $\quad \Delta H = 47.0 \text{ kJ}$

84. (a) and (d); When a process results in the evolution of heat, it is said to be exothermic (energy flows out of the system). Combustion is a chemical reaction that produces a lot of heat. Condensation releases energy to change from a gaseous state to a liquid state.

86. $Q = m \times s \times \Delta T$

$Q = 2.4 \text{ mol C} \times \dfrac{12.01 \text{ g C}}{1 \text{ mol C}} \times \dfrac{0.71 \text{ J}}{\text{g}°\text{C}} \times 25.0°\text{C} = 5.1 \times 10^2 \text{ J}$

88. $54.0 \text{ g B}_2\text{H}_6 \times \dfrac{1 \text{ mol B}_2\text{H}_6}{27.668 \text{ g B}_2\text{H}_6} \times \dfrac{2035 \text{ kJ}}{\text{mol}} = 3.97 \times 10^3 \text{ kJ}$

# CHAPTER 11

# Modern Atomic Theory

2. Rutherford was not able to determine where the electrons were in the atom or what they were doing.

4. The different forms of electromagnetic radiation are similar in that they all exhibit the same type of wave-like behavior and are propagated through space at the same speed (the speed of light). The types of electromagnetic radiation differ in their frequency (and wavelength) and in the resulting amount of energy carried per photon.

6. The *speed* of electromagnetic radiation represents how fast a given wave moves through space. The *frequency* of electromagnetic radiation represents how many complete cycles of the wave pass a given point per second. These two concepts are not the same.

8. The greenhouse gases do not absorb light in the visible wavelengths, enabling this light to pass through the atmosphere and continue to warm the earth, keeping the earth much warmer than it would be without these gases. The earth, in turn, emits infrared radiation which is absorbed by the greenhouse gases and which is re-emitted in all directions. As we increase our use of fossil fuels, the level of $CO_2$ in the atmosphere is increasing gradually, but significantly. An increase in the level of $CO_2$ will warm the earth further, eventually changing the weather patterns on the earth's surface and melting the polar ice caps.

10. exactly equal to

12. photon

14. absorbs

16. When excited hydrogen atoms emit their excess energy, the photons of radiation emitted are always of exactly the same wavelength and energy. We consider this to mean that the hydrogen atom possesses only certain allowed energy states, and that the photons emitted correspond to the atom changing from one of these allowed energy states to another of the allowed energy state. The energy of the photon emitted corresponds to the energy difference in the allowed states. If the hydrogen atom did not possess discrete energy levels, then we would expect the photons emitted to have random wavelengths and energies.

18. The energy of an emitted photon is *identical* to the energy change within the atom that gave rise to the emitted photon.

20. Energy is emitted only at wavelengths corresponding to the specific transitions for the electron among the energy levels of hydrogen.

22. The electron moves to an orbit farther from the nucleus of the atom.

Chapter 11: Modern Atomic Theory

24. Bohr's theory *explained* the experimentally *observed* line spectrum of hydrogen *exactly*. Bohr's theory was ultimately discarded because when attempts were made to extend the theory to atoms other than hydrogen, the calculated properties did *not* correspond closely to experimental measurements.

26. An orbit represents a definite, exact circular pathway around the nucleus in which an electron can be found. An orbital represents a region of space in which there is a high probability of finding the electron.

28. The firefly analogy is intended to demonstrate the concept of a probability map for electron density. In the wave mechanical model of the atom, we cannot say specifically where the electron is in the atom, we can only say where there is a high probability of finding the electron. The analogy is to imagine a time-exposure photograph of a firefly in a closed room. Most of the time, the firefly will be found near the center of the room.

30. (b)

32. The *p* orbitals, in general, have two lobes and are sometimes described as having a "dumbbell" shape. The 2*p* and 3*p* orbitals are similar in shape, and in fact there are three equivalent 2*p* or 3*p* orbitals in the 2*p* or 3*p* subshell. The orbitals differ in size, mean distance from the nucleus, and energy.

34. excited

36.  | Value of n | Possible subshells |
     |---|---|
     | 1 | 1*s* |
     | 2 | 2*s*, 2*p* |
     | 3 | 3*s*, 3*p*, 3*d* |
     | 4 | 4*s*, 4*p*, 4*d*, 4*f* |

38. Electrons have an intrinsic spin (they spin on their own axes). Geometrically, there are only two senses possible for spin (clockwise or counter-clockwise). This means only two electrons can occupy an orbital, with the opposite sense or direction of spin. This idea is called the Pauli Exclusion Principle.

40. increases; as you move out from the nucleus, there is more space and room for more sublevels.

42. opposite

44. (*a*) and (*d*); *f* orbitals begin at the fourth energy level and *p* orbitals begin at the second energy level.

46. When a hydrogen atom is in its ground state, the electron is found in the 1*s* orbital. The 1*s* orbital has the lowest energy of all the possible hydrogen orbitals.

48. The elements in a given vertical column of the periodic table have the same valence electron configuration. Having the same valence electron configuration causes the elements in a given group to have similar chemical properties.

Chapter 11: Modern Atomic Theory

50. Just count the electrons to get the atomic number of the element.

    a. silicon
    b. beryllium
    c. neon
    d. argon

52. Just count the electrons to get the atomic number of the element.

    a. selenium
    b. scandium
    c. sulfur
    d. iodine

54. a. 1s(↑↓) 2s(↑↓) 2p(↑↓)(↑↓)(↑↓) 3s(↑↓) 3p(↑ )( )( )
    b. 1s(↑↓) 2s(↑↓) 2p(↑↓)(↑↓)(↑↓) 3s(↑↓) 3p(↑ )(↑ )(↑ )
    c. 1s(↑↓) 2s(↑↓) 2p(↑↓)(↑↓)(↑↓) 3s(↑↓) 3p(↑↓)(↑↓)(↑↓) 4s(↑↓)
       3d(↑↓)(↑↓)(↑↓)(↑↓)(↑↓) 4p(↑↓)(↑↓)(↑ )

56. Specific answers depend on student choice of elements. Any Group 2 element would have two valence electrons. Any Group 4 element would have four valence electrons. Any Group 6 element would have six valence electrons. Any Group 8 element would have eight valence electrons.

58. The properties of Rb and Sr suggest that they are members of Groups 1 and 2, respectively, and so must be filling the 5s orbital. The 5s orbital is lower in energy (and fills before) the 4d orbitals.

60. a. aluminum
    b. potassium
    c. bromine
    d. tin

62. a. [Kr] $5s^1$: 1 valence electron
    b. [Ar] $4s^2\ 3d^{10}\ 4p^3$: 5 valence electrons (d electrons are not counted as valence electrons)
    c. [Ne] $3s^2\ 3p^1$: 3 valence electrons
    d. [Ar] $4s^2\ 3d^8$: 2 valence electrons (d electrons are not counted as valence electrons)

64. a. [Kr] $5s^2\ 4d^6$: 6 4d electrons
    b. [Kr] $5s^2\ 4d^8$: 8 4d electrons
    c. [Kr] $5s^2\ 4d^{10}\ 5p^2$: 10 4d electrons
    d. [Ar] $4s^2\ 3d^6$: 0 4d electrons

Chapter 11: Modern Atomic Theory

66. 
  a. $[Rn]7s^2 6d^1 5f^3$
  b. $[Ar]4s^2 3d^5$
  c. $[Xe]6s^2 4f^{14} 5d^{10}$
  d. $[Rn]7s^1$

68. $[Rn]\ 7s^2\ 5f^{14}\ 6d^5$

70. The metallic elements *lose* electrons and form *positive* ions (cations); the nonmetallic elements *gain* electrons and form *negative* ions (anions). Remember that the electron itself is *negatively* charged.

72. All exist as *diatomic* molecules ($F_2$, $Cl_2$, $Br_2$, $I_2$); all are *non*metals; all have relatively high electronegativities; all form 1- ions in reacting with metallic elements.

74. Elements at the *left* of a period (horizontal row) lose electrons more readily; at the left of a period (given principal energy level) the nuclear charge is the smallest and the electrons are least tightly held.

76. The elements of a given period (horizontal row) have valence electrons in the same principal energy level. Nuclear charge, however, increases across a period going from left to right. Atoms at the left side have smaller nuclear charges and hold onto their valence electrons less tightly.

78. When substances absorb energy the electrons become excited (move to higher energy levels). Upon returning to the ground state, energy is released, some of which is in the visible spectrum. Since we see colors, this tells us that only *certain wavelengths* of light are released, which means that only *certain transitions* are allowed. This is what is meant by quantized energy levels. If all wavelengths of light were emitted we would see white light.

80. Ionization energies decrease in going from top to bottom within a vertical group; ionization energies increase in going from left to right within a horizontal period.
  a. Li
  b. Ca
  c. Cl
  d. S

82. Atomic size increases in going from top to bottom within a vertical group; atomic size decreases in going from left to right within a horizontal period.
  a. Na
  b. S
  c. N
  d. F

84. speed of light

86. photons

Chapter 11: Modern Atomic Theory

88. quantized

90. (e); An electron's path cannot be precisely described. Orbitals describe the probabilities of finding the electron at given points in space around the nucleus.

92. a. 6 valence electrons (2 from the *s* orbital and 4 from the *p* orbital)

  b. S, Se; Both S and Se each contain 6 valence electrons with an $ns^2np^4$ configuration. Cl contains 7 valence electrons. Pb contains 4 valence electrons. And finally, Cr does not contain an $ns^2np^4$ configuration at all.

94. (d)

96. a. $1s^2\ 2s^2\ 2p^6\ 3s^2\ 3p^6\ 4s^1$  [Ar] $4s^1$

  $1s(\uparrow\downarrow)\ 2s(\uparrow\downarrow)\ 2p(\uparrow\downarrow)(\uparrow\downarrow)(\uparrow\downarrow)\ 3s(\uparrow\downarrow)\ 3p(\uparrow\downarrow)(\uparrow\downarrow)(\uparrow\downarrow)\ 4s(\uparrow\ )$

  b. $1s^2\ 2s^2\ 2p^6\ 3s^2\ 3p^6\ 4s^2\ 3d^2$  [Ar] $4s^2\ 3d^2$

  $1s(\uparrow\downarrow)\ 2s(\uparrow\downarrow)\ 2p(\uparrow\downarrow)(\uparrow\downarrow)(\uparrow\downarrow)\ 3s(\uparrow\downarrow)\ 3p(\uparrow\downarrow)(\uparrow\downarrow)(\uparrow\downarrow)\ 4s(\uparrow\downarrow)$
  $3d(\uparrow\ )(\uparrow\ )(\ )(\ )(\ )$

  c. $1s^2\ 2s^2\ 2p^6\ 3s^2\ 3p^2$  [Ne] $3s^2\ 3p^2$

  $1s(\uparrow\downarrow)\ 2s(\uparrow\downarrow)\ 2p(\uparrow\downarrow)(\uparrow\downarrow)(\uparrow\downarrow)\ 3s(\uparrow\downarrow)\ 3p(\uparrow\ )(\uparrow\ )(\ )$

  d. $1s^2\ 2s^2\ 2p^6\ 3s^2\ 3p^6\ 4s^2\ 3d^6$  [Ar] $4s^2\ 3d^6$

  $1s(\uparrow\downarrow)\ 2s(\uparrow\downarrow)\ 2p(\uparrow\downarrow)(\uparrow\downarrow)(\uparrow\downarrow)\ 3s(\uparrow\downarrow)\ 3p(\uparrow\downarrow)(\uparrow\downarrow)(\uparrow\downarrow)\ 4s(\uparrow\downarrow)\ 3d(\uparrow\downarrow)(\uparrow\ )(\uparrow\ )(\uparrow\ )(\uparrow\ )$

  e. $1s^2\ 2s^2\ 2p^6\ 3s^2\ 3p^6\ 4s^2\ 3d^{10}$  [Ar] $4s^2\ 3d^{10}$

  $1s(\uparrow\downarrow)\ 2s(\uparrow\downarrow)\ 2p(\uparrow\downarrow)(\uparrow\downarrow)(\uparrow\downarrow)\ 3s(\uparrow\downarrow)\ 3p(\uparrow\downarrow)(\uparrow\downarrow)(\uparrow\downarrow)\ 4s(\uparrow\downarrow)\ 3d(\uparrow\downarrow)(\uparrow\downarrow)(\uparrow\downarrow)(\uparrow\downarrow)(\uparrow\downarrow)$

98. a. $ns^2$
  b. $ns^2\ np^5$
  c. $ns^2\ np^4$
  d. $ns^1$
  e. $ns^2\ np^4$

100. a. $\lambda = \dfrac{h}{mv}$

  $\lambda = \dfrac{6.63 \times 10^{-34}\ \text{J s}}{(9.1 \times 10^{-31}\text{kg})[0.90 \times (3.00 \times 10^8\ \text{m s}^{-1})]}$

  $\lambda = 2.7 \times 10^{-12}$ m (0.0027 nm)

  b. $4.4 \times 10^{-34}$ m

  c. $2 \times 10^{-35}$ m

## Chapter 11: Modern Atomic Theory

The wavelengths for the ball and the person are *infinitesimally small*, whereas the wavelength for the electron is nearly the same order of magnitude as the diameter of a typical atom.

102. (d); (a) describes the outdated "plum pudding model". For (b), positively charged particles called protons *and neutrons* are found inside the nucleus. For (c), electrons located further from the nucleus have *less* predictable behavior because they contain *more* energy. For (e), quantized energy is a *discrete* spectrum, like a staircase.

104.  a.  $1s^2 2s^2 2p^4$ or $[He]2s^2 2p^4$

   b.  (↑↓)   (↑↓)   (↑↓)(↑ )(↑ )
        1s      2s       2p

   In the 2p orbital, the electrons spread out between the three different orientations to minimize repulsions (because the electrons have negative charges so they repel each other). Once each 2p orbital has one electron, the electrons pair up.

   c.  Answers will vary but at least one electron should be in a higher energy level. Example: $1s^2 2s^2 2p^3 4s^1$; Make sure there are still 8 total electrons and the orbital diagram matches the given excited state electron configuration. When an atom is in an excited state, the ground state electrons move further from the nucleus into higher energy levels.

   d.  We do not see white light because the energy that is released to return the electron to its ground state is quantized. The electron that is excited can only exist in certain energy levels, so when it is returning to a lower energy level, the energy that is released is a certain quantity, which corresponds to a specific wavelength of light (a certain color). If white light appears, then all of the colors are present which means energy is not quantized.

106.  (b) and (d) are true. Element X has chemical properties similar to those of column 6 (not column 7, the halogens). When forming an ionic compound with calcium, element X will likely form a –2 charge, giving a formula of CaX.

108.  a.  $1s^2\ 2s^2\ 2p^6\ 3s^2\ 3p^6\ 4s^2\ 3d^{10}\ 4p^5$

   b.  $1s^2\ 2s^2\ 2p^6\ 3s^2\ 3p^6\ 4s^2\ 3d^{10}\ 4p^6\ 5s^2\ 4d^{10}\ 5p^6$

   c.  $1s^2\ 2s^2\ 2p^6\ 3s^2\ 3p^6\ 4s^2\ 3d^{10}\ 4p^6\ 5s^2\ 4d^{10}\ 5p^6\ 6s^2$

   d.  $1s^2\ 2s^2\ 2p^6\ 3s^2\ 3p^6\ 4s^2\ 3d^{10}\ 4p^4$

110.  a.  five (2s, 2p)

   b.  seven (3s, 3p)

   c.  one (3s)

   d.  three (3s, 3p)

112.  F, S, Ge, Mn, Rb

114.  (a) < (c) < (b); (a) corresponds to the element Xe. (c) corresponds to the element Sb. (b) corresponds to the element In. Atomic size increases from right to left across a row on the periodic table. Xe < Sb < In

Chapter 11: Modern Atomic Theory

116. (c); The atomic size of the elements decrease going across a period from left to right because the increase in positive charge on the nucleus tends to pull the electrons closer to the nucleus.

118. Atomic size increases in going from top to bottom within a vertical group; atomic size decreases in going from left to right within a horizontal period.

   a. Ca
   b. P
   c. K

120. (b), (c), and (e)

122. a. Te
   b. Ge
   c. F (9 total electrons)

124. F and B: B is the larger atom.

   C and N: C is the larger atom.

   B and Al: Al is the larger atom.

126. 

| Electron configuration | Symbol | IE | AR |
|---|---|---|---|
| $1s^2 2s^2 2p^6 3s^2$ | Mg | 0.738 | 160 |
| $1s^2 2s^2 2p^6 3s^2 3p^4$ | S | 0.999 | 104 |
| $1s^2 2s^2 2p^6 3s^2 3p^6 4s^2$ | Ca | 0.590 | 197 |

# CHAPTER 12

# Chemical Bonding

2. bond energy

4. A covalent bond represents the *sharing* of pairs of electrons between nuclei.

6. In $H_2$ and HF, the bonding is covalent in nature, with an electron pair being shared between the atoms. In $H_2$, the two atoms are identical (the sharing is equal); in HF, the two atoms are different (the sharing is unequal) and as a result the bond is polar. Both of these are in marked contrast to the situation in NaF: NaF is an ionic compound—an electron has been completely transferred from sodium to fluorine, producing separate ions.

8. A bond is polar if the centers of positive and negative charge do not coincide at the same point. The bond has a negative end and a positive end. Polar bonds will exist in any molecule with nonidentical bonded atoms (although the molecule, as a whole, may not be polar if the bond dipoles cancel each other). Two simple examples are HF and HCl: in both cases, the negative center of charge is closer to the halogen atom.

10. The level of polarity in a polar covalent bond is determined by the difference in electronegativity of the atoms in the bond.

12.     a.    At is most electronegative, Cs is least electronegative
    b.    Sr is most electronegative, Ba and Ra have the same electronegativities
    c.    O is most electronegative, Rb is least electronegative

14. Generally, covalent bonds between atoms of *different* elements are *polar*.
    a.    covalent
    b.    polar covalent
    c.    ionic

16. For a bond to be polar covalent, the atoms involved in the bond must have different electronegativities (must be of different elements).
    a.    nonpolar covalent (atoms of the same element)
    b.    nonpolar covalent (atoms of the same element)
    c.    nonpolar covalent (atoms of the same element)
    d.    polar covalent (atoms of different elements)

Chapter 12: Chemical Bonding

18. The *degree* of polarity of a polar covalent bond is indicated by the magnitude of the difference in electronegativities of the elements involved: the larger the difference in electronegativity, the more polar the bond. Electronegativity differences are given in parentheses below:

   a. O–Cl (0.5); O–Br(0.7); the O–Br bond is more polar

   b. N–O (0.5); N–F (1.0); the N–F bond is more polar

   c. P–S (0.4); P–O (1.4); the P–O bond is more polar

   d. H–O (1.4); H–N (0.9); the H–O bond is more polar

20. The greater the electronegativity difference between two atoms, the more ionic will be the bond between those two atoms.

   a. Na–N

   b. K–P

   c. Na–Cl

   d. Mg–Cl

22. The presence of strong bond dipoles and a large overall dipole moment in water make it a polar substance overall. Among the properties of water dependent on its dipole moment are its freezing point, melting point, vapor pressure, and its ability to dissolve many substances.

24. In a diatomic molecule containing two different elements, the more electronegative atom will be the negative end of the molecule, and the *less* electronegative atom will be the positive end.

   a. H

   b. Cl

   c. I

26. In the figures, the arrow points toward the more electronegative atom.

   a. $\delta$+ P→S $\delta$–

   b. $\delta$+ S→F $\delta$–

   c. $\delta$+ S→Cl $\delta$–

   d. $\delta$+ S→Br $\delta$–

28. In the figures, the arrow points toward the more electronegative atom.

   a. $\delta$+ H→C $\delta$–

   b. $\delta$+ N→O $\delta$–

   c. $\delta$+ S→N $\delta$–

   d. $\delta$+ C→N $\delta$–

30. preceding

32. Atoms in covalent molecules gain a configuration like that of a noble gas by sharing one or more pairs of electrons between atoms: such shared pairs of electrons "belong" to each of the atoms of

## Chapter 12: Chemical Bonding

the bond at the same time. In ionic bonding, one atom completely gives over one or more electrons to another atom, and the resulting ions behave independently of one another.

34.  a. $Br^-$, Kr (Br has one electron less than Kr)
   b. $Cs^+$, Xe (Cs has one electron more than Xe)
   c. $P^{3-}$, Ar (P has three fewer electrons than Ar)
   d. $S^{2-}$, Ar (S has two fewer electrons than Ar)

36.  $[Kr]5s^2 4d^{10} 5p^6$; The ground state configuration for Te is $[Kr]5s^2 4d^{10} 5p^4$. $Te^{2-}$ contains two additional electrons, thus changing the configuration to $5p^6$.

38.  a. $AlBr_3$: Al has three electrons more than a noble gas; Br has one electron less than a noble gas.
   b. $Al_2O_3$: Al has three electrons more than a noble gas; O has two fewer electrons than a noble gas.
   c. AlP: Al has three electrons more than a noble gas; P has three fewer electrons than a noble gas.
   d. $AlH_3$: Al has three electrons more than a noble gas; H has one electron less than a noble gas.

40.  There are many examples possible. Listed below are a few compounds that fit each situation.
   a. LiF: $Li^+$, [He]; $F^-$, [Ne]
   b. NaF: $Na^+$, [Ne]; $F^-$, [Ne]
   c. LiCl: $Li^+$, [He]; $Cl^-$, [Ar]
   d. NaCl: $Na^+$, [Ne]; $Cl^-$, [Ar]

42.  An ionic solid such as NaCl consists of an array of alternating positively– and negatively–charged ions: that is, each positive ion has as its nearest neighbors a group of negative ions, and each negative ion has a group of positive ions surrounding it. In most ionic solids, the ions are packed as tightly as possible.

44.  In forming an anion, an atom gains additional electrons in its outermost (valence) shell. Additional electrons in the valence shell increases the repulsive forces between electrons, so the outermost shell becomes larger to accommodate this.

46.  Relative ionic sizes are given in Figure 12.9. Within a given horizontal row of the periodic chart, negative ions tend to be larger than positive ions because the negative ions contain a larger number of electrons in the valence shell. Within a vertical group of the periodic table, ionic size increases from top to bottom. In general, positive ions are smaller than the atoms they come from, whereas negative ions are larger than the atoms they come from. If two ions contain the same number of electrons, look at the nuclear charge. In general, the larger the nuclear charge, the smaller the ion (electrons drawn closer to the nucleus).
   a. Mg
   b. $K^+$

c. Br⁻
d. Se²⁻

48. Relative ionic sizes are given in Figure 12.9. Within a given horizontal row of the periodic chart, negative ions tend to be larger than positive ions because the negative ions contain a larger number of electrons in the valence shell. Within a vertical group of the periodic table, ionic size increases from top to bottom. In general, positive ions are smaller than the atoms they come from, whereas negative ions are larger than the atoms they come from.

   a.   I
   b.   F⁻
   c.   F⁻

50. When atoms form covalent bonds, they try to attain a valence electronic configuration similar to that of the following noble gas element. When the elements in the first few horizontal rows of the periodic table form covalent bonds, they will attempt to gain configurations similar to the noble gases helium (2 valence electrons, duet rule), and neon and argon (8 valence electrons, octet rule).

52. These elements attain a total of eight valence electrons, making the valence electron configurations similar to those of the noble gases Ne and Ar.

54. When two atoms in a molecule are connected by a triple bond, the atoms share three pairs of electrons (6 electrons) in completing their outermost shells. A simple molecule containing a triple bond is acetylene, $C_2H_2$ (H:C:::C:H).

56. The Group in which a representative element is found indicates the number of valence electrons.

   a.   ·Mg·

   b.   :B̈r·

   c.   :S̈·

   d.   ·Si·

58. a.   each boron provides 3; each oxygen provides 6; total valence electrons = 24
    b.   carbon provides 4; each oxygen provides 6; total valence electrons = 16
    c.   each carbon provides 4; each hydrogen provides 1; oxygen provides 6; total valence electrons = 20
    d.   N provides 5; each oxygen provides 6; total valence electrons = 17

60. a.   Each hydrogen provides 1 valence electron; sulfur provides 6 valence electrons; total valence electrons = 8

Chapter 12: Chemical Bonding

b. Each fluorine provides 7 valence electrons; silicon provides 4 valence electrons; total valence electrons = 32

[Lewis structure of SiF$_4$: central Si bonded to four F atoms, each F with three lone pairs]

c. Each carbon provides 4 valence electrons; each hydrogen provides 1 valence electron; total valence electrons = 12

[Lewis structure of C$_2$H$_4$: H$_2$C=CH$_2$]

d. Each carbon provides 4 valence electrons; each hydrogen provides 1 valence electron; total valence electrons = 20

[Lewis structure of C$_2$H$_6$: H$_3$C—CH$_3$]

62. NO$_2^-$ exhibits resonance. A molecule shows resonance when more than one Lewis structure can be drawn for the molecule.

[Lewis structure of CH$_4$: central C bonded to four H]

[Lewis structure of Cl$_2$O: Cl—O—Cl with lone pairs]

[Resonance structures of NO$_2^-$: [O=N—O]$^-$ ↔ [O—N=O]$^-$]

[Lewis structure of HCN: H—C≡N:]

64. C provides 4 valence electrons. Each oxygen provides 6 valence electrons. Having only 16 total valence electrons requires multiple bonding in the molecule.

:O≡C—Ö:  ↔  Ö=C=Ö  ↔  :Ö—C≡O:

Chapter 12: Chemical Bonding

66. a. Cl provides 7 valence electrons. Each O provides 6 valence electrons. The 1– charge means 1 additional electron. Total valence electrons = 26

$$\left[ \begin{array}{c} :\ddot{O}: \\ | \\ :\ddot{O}-\overset{}{Cl}-\ddot{O}: \end{array} \right]^{1-}$$

b. Each O provides 6 valence electrons. The 2– charge means two additional valence electrons. Total valence electrons = 14

$$\left[ :\ddot{O}-\ddot{O}: \right]^{2-}$$

c. Each C provides 4 valence electrons. Each H provides 1 valence electron. Each O provides 6 valence electrons. The 1– charge means 1 additional valence electron. Total valence electrons = 24

$$\left[ \begin{array}{c} H \quad :O: \\ | \quad \| \\ H-C-C-\ddot{O}: \\ | \\ H \end{array} \right]^{1-} \quad \left[ \begin{array}{c} H \quad :\ddot{O}: \\ | \quad | \\ H-C-C=\ddot{O} \\ | \\ H \end{array} \right]^{1-}$$

68. a. C provides 4 valence electrons. Each O provides 6 valence electrons. The 2– charge means two additional valence electrons.
Total valence electrons = 24

$$\left[ \begin{array}{ccc} :O: & :\ddot{O}: & :\ddot{O}: \\ \| & | & | \\ C & \leftrightarrow \quad C \quad \leftrightarrow & C \\ /\backslash & /\backslash & /\backslash \\ :\ddot{O}: \quad :\ddot{O}: & :\ddot{O}: \quad :\ddot{O}: & :\ddot{O}: \quad :\ddot{O}: \end{array} \right]^{2-}$$

b. Each H provides 1 valence electron. N provides 5 valence electrons. The 1+ charge means one less valence electron.
Total valence electrons = 8

$$\left[ \begin{array}{c} H \\ | \\ H-N-H \\ | \\ H \end{array} \right]^{+}$$

c. Cl provides 7 valence electrons. O provides 6 valence electrons. The 1– charge means one additional valence electron. Total valence electrons = 14

$$\left[ :\ddot{\underset{..}{Cl}}-\ddot{\underset{..}{O}}: \right]^{-}$$

70. The geometric structure of NH$_3$ is that of a trigonal pyramid. The nitrogen atom of NH$_3$ is surrounded by four electron pairs (three are bonding, one is a lone pair). The H–N–H bond angle is somewhat less than 109.5° (due to the presence of the lone pair).

## Chapter 12: Chemical Bonding

72. The geometric structure of $SiF_4$ is that of a tetrahedron. The silicon atom of $SiF_4$ is surrounded by four bonding electron pairs. The F–Si–F bond angle is the characteristic angle of the tetrahedron, 109.5°.

74. The general molecular structure of a molecule is determined by (1) *how many electron pairs* surround the central atom in the molecule, and (2) which of those electron pairs are used for *bonding* to the other atoms of the molecule. Nonbonding electron pairs on the central atom do, however, cause minor changes in the bond angles, compared to the ideal regular geometric structure.

76. You will remember from high school geometry, that two points in space are all that is needed to define a straight line. A diatomic molecule represents two points (the nuclei of the atoms) in space.

78. In $NF_3$, the nitrogen atom has *four* pairs of valence electrons, whereas in $BF_3$, there are only *three* pairs of valence electrons around the boron atom. The nonbonding electron pair on nitrogen in $NF_3$ pushes the three F atoms out of the plane of the N atom.

80. a. four electron pairs in a tetrahedral arrangement with some lone-pair distortion
    b. four electron pairs in a tetrahedral arrangement with some lone-pair distortion
    c. four electron pairs in a tetrahedral arrangement

82. a. tetrahedral
    b. trigonal pyramidal (there is a lone pair on P)
    c. non-linear, *V*-shaped (four electron pairs on O, but only two atoms are attached to O)

84. a. basically tetrahedral around the P atom (the hydrogen atoms are attached to two of the oxygen atoms and do not affect greatly the geometrical arrangement of the oxygen atoms around the phosphorus)
    b. tetrahedral (4 electron pairs on Cl, and 4 atoms attached)
    c. trigonal pyramidal (4 electron pairs on S, and 3 atoms attached)

86. a. approximately 109.5° (the molecule is *V*-shaped or nonlinear)
    b. approximately 109.5° (the molecule is trigonal pyramidal)
    c. 109.5°
    d. approximately 120° (the double bond makes the molecule flat)

88. a. *V*-shaped; 120° (one lone pair on Se with two S atoms attached, one with a double bond)
    b. trigonal planar; 120° (three S atoms attached to Se, one with a double bond)
    c. *V*-shaped; 120° (one lone pair on S with two O atoms attached, one with a double bond)
    d. linear; 180° (two S atoms attached to C, each with a double bond)

90. (*d*); In covalent bonding, the electrons are shared between the atoms. The process of forming an ionic bond is highly exothermic overall. The bonding that occurs in ionic bonding is usually between metal and nonmetal atoms.

92. C–F, Si–F, Ga–F, Na–F; The greater the difference in electronegativity values between the two atoms, the more polar the bond.

94. (*e*); The bond in KBr is formed by ionic bonding, in which an electron is transferred from K to Br.

96. In each case, the element *higher up* within a group on the periodic table has the higher electronegativity.

   a. Be
   b. N
   c. F

98. For a bond to be polar covalent, the atoms involved in the bond must have different electronegativities (must be of different elements).

   a. polar covalent (different elements)
   b. *non*polar covalent (two atoms of the same element)
   c. polar covalent (different elements)
   d. *non*polar covalent (atoms of the same element)

100. (*b*); Fluorine has a better ability to attract shared electrons to itself.

102. AgCl; Ag forms a +1 charge to form an ionic bond with Cl (which becomes Cl$^-$). Thus Ag$^+$ contains the same number of electrons as Pd, which is not a noble gas.

104. a. Na$_2$Se: Na has one electron more than a noble gas; Se has two electrons fewer than a noble gas.
   b. RbF: Rb has one electron more than a noble gas; F has one electron less than a noble gas.
   c. K$_2$Te: K has one electron more than a noble gas; Te has two electrons fewer than a noble gas.
   d. BaSe: Ba has two electrons more than a noble gas; Se has two electrons fewer than a noble gas.
   e. KAt: K has one electron more than a noble gas; At has one electron less than a noble gas.
   f. FrCl: Fr has one electron more than a noble gas; Cl has one electron less than a noble gas.

106. Relative ionic sizes are indicated in Figure 12.9.

   a. Na$^+$
   b. Al$^{3+}$
   c. F$^-$
   d. Na$^+$

Chapter 12: Chemical Bonding

108.   a.   H provides 1; N provides 5; each O provides 6; total valence electrons = 24
       b.   each H provides 1; S provides 6; each O provides 6; total valence electrons = 32
       c.   each H provides 1; P provides 5; each O provides 6; total valence electrons = 32
       d.   H provides 1; Cl provides 7; each O provides 6; total valence electrons = 32

110.   (d); The Lewis structure given contains 14 valence electrons. This matches $O_2^{2-}$: ($6e^- + 6e^- + 2e^-$ from the charge = 14 total valence electrons).

112.   $H_2O$, HCl; Both $H_2O$ and HCl are polar molecules and thus contain a net dipole moment.

114.   a.   four electron pairs arranged tetrahedrally about C
       b.   four electron pairs arranged tetrahedrally about Ge
       c.   three electron pairs arranged trigonally (planar) around B

116.   a.   $ClO_3^-$, trigonal pyramid (lone pair on Cl)
       b.   $ClO_2^-$, nonlinear (V-shaped, two lone pairs on Cl)
       c.   $ClO_4^-$, tetrahedral (all pairs on Cl are bonding)

118.   a.   nonlinear (V–shaped)
       b.   trigonal planar
       c.   basically trigonal planar around the C (the H is attached to one of the O atoms, and distorts the shape around the carbon only slightly)
       d.   linear

120.   $[Kr]5s^24d^{10}5p^5$; The two elements with one unpaired 5p electron are indium (In) and iodine (I), however indium would form an ionic compound with fluorine because of it's more metallic nature. Thus, the element must be iodine, which is a nonmetal, and thus forms a covalent compound with fluorine.

122.   Generally, covalent bonds between atoms of *different* elements are *polar*.
       a.   nonpolar covalent
       b.   ionic
       c.   ionic
       d.   polar covalent
       e.   polar covalent
       f.   nonpolar covalent
       g.   nonpolar covalent
       h.   ionic

124.   O–F, P–Cl, P–F, Si–F; The greater the electronegativity difference between two atoms, the more polar the bond between those two atoms.

Chapter 12: Chemical Bonding

126. Na$^+$: $1s^2\,2s^2\,2p^6$

K$^+$: $1s^2\,2s^2\,2p^6\,3s^2\,3p^6$

Li$^+$: $1s^2$

Cs$^+$: $1s^2\,2s^2\,2p^6\,3s^2\,3p^6\,4s^2\,3d^{10}\,4p^6\,5s^2\,4d^{10}\,5p^6$

128.

| Formula | Compound Name | Molecular Structure |
|---|---|---|
| $CO_2$ | carbon dioxide | linear |
| $NH_3$ | ammonia | trigonal pyramidal |
| $SO_3$ | sulfur trioxide | trigonal planar |
| $H_2O$ | water | bent or V-shaped |
| $ClO_4^-$ | perchlorate ion | tetrahedral |

# CUMULATIVE REVIEW

# Chapters 10–12

2. Temperature is a measure of the random motions of the components of a substance: in other words, temperature is a measure of the average kinetic energy of the particles in a sample. The molecules in warm water must be moving faster than the molecules in cold water (the molecules have the same mass, so if the temperature is higher, the average velocity of the particles must be higher in the warm water). Heat is the energy that flows because of a difference in temperature.

4. Thermodynamics is the study of energy and energy changes. The first law of thermodynamics is the law of conservation of energy: the energy of the universe is constant. Energy cannot be created or destroyed, only transferred from one place to another or from one form to another. The internal energy of a system, $E$, represents the total of the kinetic and potential energies of all the particles in a system. A flow of heat may be produced when there is a change in internal energy in the system, but it is not correct to say that the system "contains" the heat: part of the internal energy is *converted* to heat energy during the process (under other conditions, the change in internal energy might be expressed as work rather than a heat flow).

6. The enthalpy change represents the heat energy that flows (at constant pressure) on a molar basis when a reaction occurs. The enthalpy change is indeed a state function (which we make great use of in Hess's Law calculations). Enthalpy changes are typically measured in insulated reaction vessels called calorimeters (a simple calorimeter is shown in Figure 10.6 in the text).

8. Consider petroleum. A gallon of gasoline contains concentrated, stored energy. We can use that energy to make our car move, but when we do, the energy stored in the gasoline is dispersed to the environment. Although the energy is still there (it is conserved), it is no longer in a concentrated useful form. So although the energy content of the universe remains constant, the energy that is now stored in concentrated forms in oil, coal, wood, and other sources is gradually being dispersed to the universe where it can do no work.

10. A "driving force" is an effect that tends to make a process occur. Two important driving forces are dispersion of energy during a process or dispersion of matter during a process ("energy spread" and "matter spread"). For example, a log burns in a fireplace because the energy contained in the log is dispersed to the universe when it burns. If we put a teaspoon of sugar into a glass of water, the dissolving of the sugar is a favorable process because the matter of the sugar is dispersed when it dissolves. Entropy is a measure of the randomness or disorder in a system. The entropy of the universe is constant increasing because of "matter spread" and "energy spread". A spontaneous process is one that occurs without outside intervention: the spontaneity of a reaction depends on the energy spread and matter spread if the reaction takes place. A reaction that disperses energy and also disperses matter will always be spontaneous. Reactions that require an input of energy may still be spontaneous if the matter spread is large enough.

12.    molar mass CH$_4$ = 16.04 g

　a.　　$0.521 \text{ mol} \times \dfrac{-890 \text{ kJ}}{1 \text{ mol}} = -464 \text{ kJ}$

　b.　　$1.25 \text{ g} \times \dfrac{1 \text{ mol}}{16.04 \text{ g}} \times \dfrac{-890 \text{ kJ}}{1 \text{ mol}} = -69.4 \text{ kJ}$

　c.　　$-1250 \text{ kJ} \times \dfrac{1 \text{ mol}}{-890 \text{ kJ}} = 1.40 \text{ mol } (22.5 \text{ g})$

14.    An atom is said to be in its ground state when it is in its lowest possible energy state. When an atom possesses more energy than its ground state energy, the atom is said to be in an excited state. An atom is promoted from its ground state to an excited state by absorbing energy; when the atom returns from an excited state to its ground state it emits the excess energy as electromagnetic radiation. Atoms do not gain or emit radiation randomly, but rather do so only in discrete bundles of radiation called photons. The photons of radiation emitted by atoms are characterized by the wavelength (color) of the radiation: longer wavelength photons carry less energy than shorter wavelength photons. The energy of a photon emitted by an atom corresponds exactly to the difference in energy between two allowed energy states in an atom: thus, we can use an observable phenomenon (emission of light by excited atoms), to gain insight into the energy changes taking place within the atom.

16.    Bohr pictured the electron moving in only certain circular orbits around the nucleus. Each particular orbit (corresponding to a particular distance from the nucleus) had associated with it a particular energy (resulting from the attraction between the nucleus and the electron). When an atom absorbs energy, the electron moves from its ground state in the orbit closest to the nucleus ($n = 1$) to an orbit farther away from the nucleus ($n = 2, 3, 4, ...$). When an excited atom returns to its ground state, corresponding to the electron moving from an outer orbit to the orbit nearest the nucleus, the atom emits the excess energy as radiation. As the Bohr orbits are of fixed distances from the nucleus and from each other, when an electron moves from one fixed orbit to another, the energy change is of a definite amount. This corresponds to a photon being emitted of a particular characteristic wavelength and energy. The original Bohr theory worked very well for hydrogen: Bohr even predicted emission wavelengths for hydrogen that had not yet been seen, but were subsequently found at the exact wavelengths Bohr had calculated. However, when the simple Bohr model for the atom was applied to the emission spectra of other elements, the theory could not predict or explain the observed emission spectra.

18.    The lowest energy hydrogen atomic orbital is called the 1s orbital. The 1s orbital is spherical in shape (that is, the electron density around the nucleus is uniform in all directions from the nucleus). The 1s orbital represents a probability map of electron density around the nucleus for the first principal energy level. The orbital does not have a sharp edge (it appears fuzzy) because the probability of finding the electron does not drop off suddenly with distance from the nucleus. The orbital does not represent just a spherical surface on which the electron moves (this would be similar to Bohr's original theory). When we draw a picture to represent the 1s orbital we are indicating that the probability of finding the electron within this region of space is greater than 90%. We know that the likelihood of finding the electron within this orbital is very high, but we still don't know exactly where in this region the electron is at a given instant in time.

Review:   Chapters 10, 11, and 12

20. The third principal energy level of hydrogen is divided into three sublevels, the 3s, 3p, and 3d sublevels. The 3s subshell consists of the single 3s orbital: like the other s orbitals, the 3s orbital is spherical in shape. The 3p subshell consists of a set of three equal-energy 3p orbitals: each of these 3p orbitals has the same shape ("dumbbell"), but each of the 3p orbitals is oriented in a different direction in space. The 3d subshell consists of a set of five 3d orbitals: the 3d orbitals have the shapes indicated in Figure 11.28, and are oriented in different directions around the nucleus (students sometimes say that the 3d orbitals have the shape of a 4-leaf clover). The fourth principal energy level of hydrogen is divided into four sublevels, the 4s, 4p, 4d, and 4f orbitals. The 4s subshell consists of the single 4s orbital. The 4p subshell consists of a set of three 4p orbitals. The 4d subshell consists of a set of five 4d orbitals. The shapes of the 4s, 4p, and 4d orbitals are the same as the shapes of the orbitals of the third principal energy level (the orbitals of the fourth principal energy level are larger and further from the nucleus than the orbitals of the third level, however). The fourth principal energy level, because it is further from the nucleus, also contains a 4f subshell, consisting of seven 4f orbitals (the shapes of the 4f orbitals are beyond the scope of this text).

22. Atoms have a series of principal energy levels symbolized by the letter $n$. The $n = 1$ level is the closest to the nucleus, and the energies of the levels increase as the value of $n$ increases going out from the nucleus. Each principal energy level is divided into a set of sublevels of different characteristic shapes (designated by the letters s, p, d, and f). Each sublevel is further subdivided into a set of orbitals: each s subshell consists of a single s orbital; each p subshell consists of a set of three p orbitals; each d subshell consists of a set of five d orbitals; etc. A given orbital can be empty or it can contain one or two electrons, but never more than two electrons (if an orbital contains two electrons, then the electrons must have opposite intrinsic spins). The shape we picture for an orbital represents only a probability map for finding electrons: the shape does not represent a trajectory or pathway for electron movements.

24. The valence electrons are the electrons in an atom's outermost shell. The valence electrons are those most likely to be involved in chemical reactions because they are at the outside edge of the atom.

26. From the column and row location of an element, you should be able to determine what the valence shell of an element has for its electronic configuration. For example, the element in the third horizontal row, in the second vertical column, has $3s^2$ as its valence configuration. We know that the valence electrons are in the $n = 3$ shell because the element is in the third horizontal row. We know that the valence electrons are s electrons because the first two electrons in a horizontal row are always in an s subshell. We know that there are two electrons because the element is the second element in the horizontal row. As an additional example, the element in the seventh vertical column of the second horizontal row in the periodic table has valence configuration $2s^2 2p^5$.

28. The ionization energy of an atom represents the energy required to remove an electron from the atom. As one goes from top to bottom in a vertical group in the periodic table, the ionization energies decrease (it becomes easier to remove an electron). As one goes down within a group, the valence electrons are farther and farther from the nucleus and are less tightly held. The ionization energies increase when going from left to right within a horizontal row within the periodic table. The left-hand side of the periodic table is where the metallic elements are found, which lose electrons relatively easily. The right-hand side of the periodic table is where the nonmetallic elements are found: rather than losing electrons, these elements tend to gain electrons. Within a given horizontal row in the periodic table, the valence electrons are all in the same principal energy shell: however, as you go from left to right in the horizontal row, the

nuclear charge that holds onto the electrons is increasing one unit with each successive element, making it that much more difficult to remove an electron. The relative sizes of atoms also vary systematically with the location of an element in the periodic table. Within a given vertical group, the atoms get progressively larger when going from the top of the group to the bottom: the valence electrons of the atoms are in progressively higher principal energy shells (and are progressively further from the nucleus) as we go down in a group. In going from left to right within a horizontal row in the periodic table, the atoms get progressively smaller. Although all the elements in a given horizontal row in the periodic table have their valence electrons in the same principal energy shell, the nuclear charge is progressively increasing from left to right, making the given valence shell progressively smaller as the electrons are drawn more closely to the nucleus.

30. Ionic bonding results when elements of very different electronegativities react with each other. Typically a metallic element reacts with a nonmetallic element; the metallic element losing electrons and forming positive ions and the nonmetallic element gaining electrons and forming negative ions. Sodium chloride, NaCl, is an example of a typical ionic compound. The aggregate form of such a compound consists of a crystal lattice of alternating positively and negatively charged ions. A given positive ion is attracted by several surrounding negatively charged ions, and a given negative ion is attracted by several surrounding positively charged ions. Similar electrostatic attractions go on in three dimensions throughout the crystal of ionic solid, leading to a very stable system (with very high melting and boiling points, for example). We know that ionic-bonded solids do not conduct electricity in the solid state (because the ions are held tightly in place by all the attractive forces), but such substances are strong electrolytes when melted or when dissolved in water (either process sets the ions free to move around).

32. Electronegativity represents the relative ability of an atom in a molecule to attract shared electrons towards itself. In order for a bond to be polar, one of the atoms in the bond must attract the shared electron pair towards itself and away from the other atom of the bond: this can only happen if one atom of the bond is more electronegative than the other (that is, that there is a considerable difference in electronegativity for the two atoms of the bond). The larger the difference in electronegativity between two atoms joined in a bond, the more polar is the bond. Specific examples depend on student choice of elements, but in general, a molecule like $Cl_2$ would be non-polar because both atoms of the bond have the same electronegativity, whereas a molecule like HCl would be polar because there is an electronegativity difference between the two atoms in the bond.

34. It has been observed over many experiments that when an active metal like sodium or magnesium reacts with a nonmetal, the sodium atoms always form $Na^+$ ions and the magnesium atoms always form $Mg^{2+}$ ions. It has been further observed that aluminum always forms only the $Al^{3+}$ ion. When nitrogen, oxygen, or fluorine form simple ions, the ions that are formed are always $N^{3-}$, $O^{2-}$, and $F^-$, respectively. Clearly the facts that these elements always form the same ions and that those ions all contain eight electrons in the outermost shell, led scientists to speculate that there must be something fundamentally stable about a species that has eight electrons in its outermost shell (like the noble gas neon). The repeated observation that so many elements, when reacting, tend to attain an electronic configuration that is isoelectronic with a noble gas led chemists to speculate that all elements try to attain such a configuration for their outermost shells. In general, when atoms of a metal react with atoms of a nonmetal, the metal atoms lose electrons until they have the configuration of the preceding noble gas, and the nonmetal atoms gain electrons until they have the configuration of the following noble gas. Covalently and polar covalently bonded molecules also strive to attain pseudo-noble gas electronic configurations. For a covalently bonded molecule like $F_2$, in which neither fluorine atom has a greater tendency than the other to

Review: Chapters 10, 11, and 12

gain or lose electrons completely, each F atom provides one electron of the pair of electrons that constitutes the covalent bond. Each F atom feels also the influence of the other F atom's electron in the shared pair, and each F atom effectively fills its outermost shell. Similarly, in polar covalently bonded molecules like HF or HCl, the shared pair of electrons between the atoms effectively completes the outer electron shell of each atom simultaneously to give each atom a noble gas-like electronic configuration.

36. Bonding between atoms to form a molecule involves only the valence electrons of the atoms (not the inner core electrons). So when we draw the Lewis structure of a molecule, we show only these valence electrons (both bonding valence electrons and nonbonding valence electrons, however). The most important requisite for the formation of a stable compound (which we try to demonstrate when we write Lewis structures) is that each atom of a molecule attains a noble gas electron configuration. When we write Lewis structures, we arrange the bonding and nonbonding valence electrons to try to complete the octet (or duet) for as many atoms as is possible.

38. Obviously, you could choose practically any molecule for your discussion. Let's illustrate the method for ammonia, $NH_3$. First count up the total number of valence electrons available in the molecule (without regard to what atom they officially come from); remember that for the representative elements, the number of valence electrons is indicated by what group the element is found in on the periodic table. For $NH_3$, because nitrogen is in Group 5, one nitrogen atom would contribute five valence electrons. Because hydrogen atoms only have one electron each, the three hydrogen atoms provide an additional three valence electrons, for a total of eight valence electrons overall. Next write down the symbols for the atoms in the molecule, and use one pair of electrons (represented by a line) to form a bond between each pair of bound atoms.

$$\begin{array}{c} H-N-H \\ | \\ H \end{array}$$

These three bonds use six of the eight valence electrons. Since each hydrogen already has its duet in what we have drawn so far, while the nitrogen atom only has six electrons around it so far, the final two valence electrons must represent a lone pair on the nitrogen.

$$\begin{array}{c} \ddot{} \\ H-N-H \\ | \\ H \end{array}$$

40. There are several types of exceptions to the octet rule described in the text. The octet rule is really a "rule of thumb" which we apply to molecules unless we have some evidence that a molecule does not follow the rule. There are some common molecules that, from experimental measurements, we know do not follow the octet rule. Boron and beryllium compounds sometimes do not fit the octet rule. For example, in $BF_3$, the boron atom only has six valence electrons, whereas in $BeF_2$, the beryllium atom only has four valence electrons. Other molecules that are exceptions to the octet rule include any molecule with an odd number of valence electrons (such as NO or $NO_2$): you can't get an octet (an even number) of electrons around each atom in a molecule with an odd number of valence electrons. Even the oxygen gas we breathe is an exception to the octet rule: although we can write a Lewis structure for $O_2$ satisfying the octet rule for each oxygen, we know from experiment that $O_2$ contains unpaired electrons (which would not be consistent with a structure in which all the electrons were paired up.)

Review: Chapters 10, 11, and 12

42. a. 6 valence electrons; 2 electrons from the s orbital and 4 electrons from the p orbital

b. S, Se; The electron configuration for S is [Ne] $3s^2\ 3p^4$ and for Se is [Ar] $4s^2\ 3d^{10}\ 4p^4$. The other elements do not have an $ns^2np^4$ configuration.

c. $K_2X$; To create an ionic compound with potassium, potassium would give up one electron to form $K^+$ (with a noble gas configuration of Ar) and the nonmetal element would gain two electrons to form $X^{2-}$ (with a noble gas configuration containing 8 valence electrons). Thus, two $K^+$ ions are needed to balance the 2– charge on X.

44. a. [Kr] $5s^2$
b. [Ne] $3s^2\ 3p^1$
c. [Ne] $3s^2\ 3p^5$
d. [Ar] $4s^1$
e. [Ne] $3s^2\ 3p^4$
f. [Ar] $4s^2\ 3d^{10}\ 4p^3$

46.

| Structure | Description |
|---|---|
| 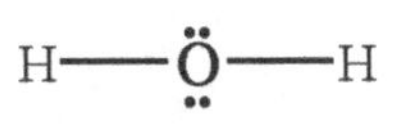 H—Ö—H | 4 electron pairs tetrahedrally-oriented on O; non-linear (bent, V-shaped) geometry; H–O–H bond angle slightly less than 109.5° because of lone pairs. |
| 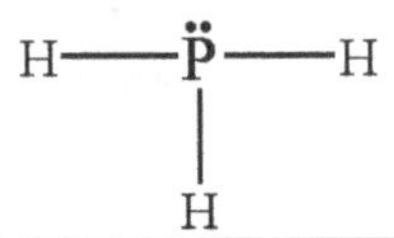 H—P̈—H, H | 4 electron pairs tetrahedrally-oriented on P; trigonal pyramidal geometry; H–P–H bond angles slightly less than 109.5° because of lone pair. |
| 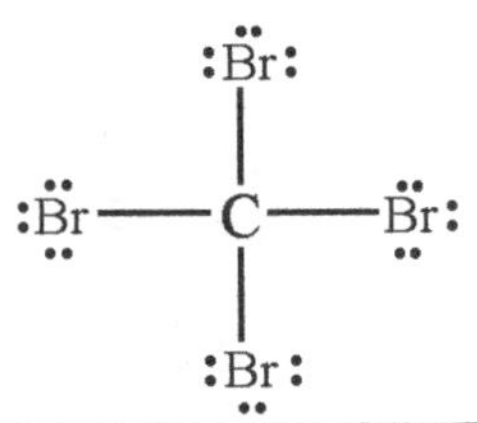 :Br̈: / :B̈r—C—B̈r: / :B̈r: | 4 electron pairs tetrahedrally-oriented on C; overall tetrahedral geometry; Br–C–Br bond angles 109.5° |
| 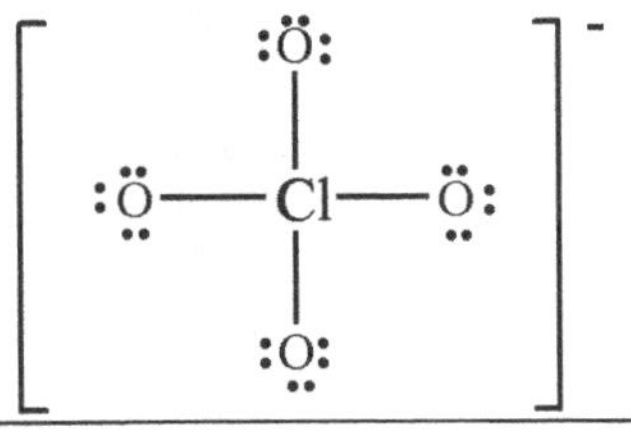 [ :Ö: / :Ö—Cl—Ö: / :Ö: ]⁻ | 4 electron pairs tetrahedrally-oriented on Cl; overall tetrahedral geometry; O–Cl–O bond angles 109.5° |
| 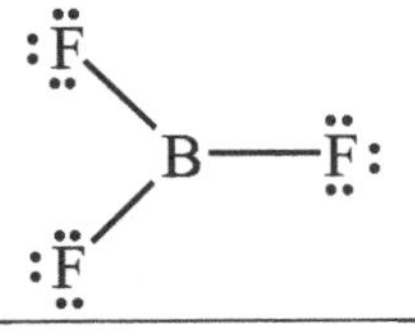 :F̈\ B—F̈: / :F̈/ | 3 electron pairs trigonally-oriented on B (exception to octet rule); overall trigonal geometry; F–B–F bond angles 120° |
| 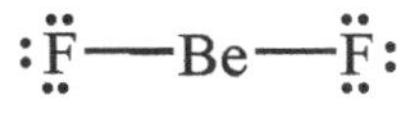 :F̈—Be—F̈: | 2 electron pairs linearly-oriented on Be (exception to octet rule); overall linear geometry; F–Be–F bond angle 180°. |

Review: Chapters 10, 11, and 12

48.  a.  H—H    :Ö=Ö:    H—Ö—H (with lone pairs on O)

  b.  H₂: Nonpolar; Equal sharing of electrons; no net dipole moment

  O₂: Nonpolar; Equal sharing of electrons; no net dipole moment

  H₂O: Polar; Unequal sharing of electrons; net dipole moment

  The polarity of water molecules causes them to attract each other strongly. This means that much energy is required to change water from a liquid to a gas. Therefore, it is the polarity of the water molecule that causes water to remain a liquid at room temperature.

  c.  Water is lower in energy for this reaction. It takes more energy to break the bonds in water than it does to break the bonds in hydrogen and oxygen. A lot more energy is given off when the water is formed (in the form of a loud noise) than what was required to break the initial hydrogen and oxygen bonds. This is an exothermic reaction and thus the product (which is water) is more stable than the reactants (hydrogen and oxygen) and is thus lower in energy.

# CHAPTER 13

# Gases

2. Solids and liquids have essentially fixed volumes and are not able to be compressed easily. Gases have volumes that depend on their conditions, and can be compressed or expanded by changes in those conditions. Although the particles of matter in solids are essentially fixed in position (the solid is rigid), the particles in liquids and gases are free to move.

4. Figure 13.2 in the text shows a simple mercury barometer: a tube filled with mercury is inverted over a reservoir (containing mercury) that is open to the atmosphere. When the tube is inverted, the mercury falls to a level at which the pressure of the atmosphere is sufficient to support the column of mercury. One standard atmosphere of pressure is taken to be the pressure capable of supporting a column of mercury to a height of 760.0 mm above the reservoir level.

6. Pressure units include mm Hg, torr, pascals, and psi. The unit "mm Hg" is derived from the barometer, since in a traditional mercury barometer, we measure the height of the mercury column (in millimeters) above the reservoir of mercury.

8. 1.00 atm = 760 torr = 760 mm Hg = 101.325 kPa = 14.70 psi

   a. $14.9 \text{ psi} \times \dfrac{1 \text{ atm}}{14.70 \text{ psi}} = 1.01 \text{ atm}$

   b. $795 \text{ torr} \times \dfrac{1 \text{ atm}}{760 \text{ torr}} = 1.05 \text{ atm}$

   c. $743 \text{ mm Hg} \times \dfrac{101.325 \text{ kPa}}{760 \text{ mm Hg}} = 99.1 \text{ kPa}$

   d. $99{,}436 \text{ Pa} \times \dfrac{1 \text{ kPa}}{1000 \text{ Pa}} = 99.436 \text{ kPa}$

10. 1.00 atm = 760 torr = 760 mm Hg = 101.325 kPa = 14.70 psi

    a. $17.3 \text{ psi} \times \dfrac{101.325 \text{ kPa}}{14.70 \text{ psi}} = 119 \text{ kPa}$

    b. $1.15 \text{ atm} \times \dfrac{14.70 \text{ psi}}{1 \text{ atm}} = 16.9 \text{ psi}$

    c. $4.25 \text{ atm} \times \dfrac{760 \text{ mm Hg}}{1 \text{ atm}} = 3.23 \times 10^3 \text{ mm Hg}$

    d. $224 \text{ psi} \times \dfrac{1 \text{ atm}}{14.70 \text{ psi}} = 15.2 \text{ atm}$

## Chapter 13: Gases

12. 1.00 atm = 760 torr = 760 mm Hg = 101,325 Pa

    a. $774 \text{ torr} \times \dfrac{1 \text{ atm}}{760 \text{ torr}} \times \dfrac{101{,}325 \text{ Pa}}{1 \text{ atm}} = 1.03 \times 10^5 \text{ Pa}$

    b. $0.965 \text{ atm} \times \dfrac{101{,}325 \text{ Pa}}{1 \text{ atm}} = 9.78 \times 10^4 \text{ Pa}$

    c. $112.5 \text{ kPa} \times \dfrac{1000 \text{ Pa}}{1 \text{ kPa}} = 1.125 \times 10^5 \text{ Pa}$

    d. $801 \text{ mm Hg} \times \dfrac{1 \text{ atm}}{760 \text{ mm Hg}} \times \dfrac{101{,}325 \text{ Pa}}{1 \text{ atm}} = 1.07 \times 10^5 \text{ Pa}$

14. Additional mercury increases the pressure on the gas sample, causing the volume of the gas upon which the pressure is exerted to decrease (Boyle's Law)

16. $PV = k$;  $P_1V_1 = P_2V_2$

18. a. $P_1 = 1.15 \text{ atm}$          $P_2 = 775 \text{ mm Hg} = 1.020 \text{ atm}$
    $V_1 = 375 \text{ mL}$          $V_2 = ?$

    $V_2 = \dfrac{P_1V_1}{P_2} = \dfrac{(1.15 \text{ atm})(375 \text{ mL})}{(1.020 \text{ atm})} = 423 \text{ mL}$

    b. $P_1 = 1.08 \text{ atm}$          $P_2 = 135 \text{ kPa} = 1.33 \text{ atm}$
    $V_1 = 195 \text{ mL}$          $V_2 = ?$

    $V_2 = \dfrac{P_1V_1}{P_2} = \dfrac{(1.08 \text{ atm})(195 \text{ mL})}{(1.33 \text{ atm})} = 158 \text{ mL}$

    c. $P_1 = 131 \text{ kPa} = 982.6 \text{ mm Hg}$          $P_2 = 765 \text{ mm Hg}$
    $V_1 = 6.75 \text{ L}$          $V_2 = ?$

    $V_2 = \dfrac{P_1V_1}{P_2} = \dfrac{(982.6 \text{ mm Hg})(6.75 \text{ L})}{(765 \text{ mm Hg})} = 8.67 \text{ L}$

20. a. $P_1 = 785 \text{ mm Hg}$          $P_2 = 700. \text{ mm Hg}$
    $V_1 = 53.2 \text{ mL}$          $V_2 = ?$

    $V_2 = \dfrac{P_1V_1}{P_2} = \dfrac{(785 \text{ mm Hg})(53.2 \text{ mL})}{700. \text{ mm Hg}} = 59.7 \text{ mL}$

    b. $P_1 = 1.67 \text{ atm}$          $P_2 = ?$
    $V_1 = 2.25 \text{ L}$          $V_2 = 2.00 \text{ L}$

    $P_2 = \dfrac{P_1V_1}{V_2} = \dfrac{(1.67 \text{ atm})(2.25 \text{ L})}{2.00 \text{ L}} = 1.88 \text{ atm}$

Chapter 13: Gases

c.  $P_1 = 695$ mm Hg  $\qquad\qquad P_2 = 1.51$ atm

$V_1 = 5.62$ L  $\qquad\qquad V_2 = ?$

$$V_2 = \frac{P_1V_1}{P_2} = \frac{(695 \text{ mm Hg})(5.62 \text{ L})}{1.51 \text{ atm} \times \left(\frac{760 \text{ mm Hg}}{1 \text{ atm}}\right)} = 3.40 \text{ L}$$

22.  $P_1 = P_1$  $\qquad\qquad P_2 = 2 \times P_1$

$V_1 = 1.04$ L  $\qquad\qquad V_2 = ?$ L

$$V_2 = \frac{P_1V_1}{P_2} = \frac{(P_1)(1.04 \text{ L})}{(2 \times P_1)} = \frac{1.04 \text{ L}}{2} = 0.520 \text{ L}$$

24.  $$P_2 = \frac{P_1V_1}{V_2} = \frac{\left(760. \text{ mm Hg} \times \frac{1 \text{ atm}}{760 \text{ mm Hg}}\right)(1.00 \text{ L})}{\left(50.0 \text{ mL} \times \frac{1 \text{ L}}{1000 \text{ mL}}\right)} = 20.0 \text{ atm}$$

26.  Charles's Law indicates that an ideal gas decreases by 1/273 of its volume for every degree Celsius its temperature is lowered. This means an ideal gas would approach a volume of zero at –273°C.

28.  $V = kT$;  $V_1/T_1 = V_2/T_2$

30.  $V_1 = 375$ mL  $\qquad\qquad V_2 = ?$ mL

$T_1 = 78$°C $= 351$ K  $\qquad\qquad T_2 = 22$°C $= 295$ K

$$V_2 = \frac{V_1T_2}{T_1} = \frac{(375 \text{ mL})(295 \text{ K})}{(351 \text{ K})} = 315 \text{ mL}$$

32.  a.  $V_1 = 25.0$ L  $\qquad\qquad V_2 = 50.0$ L

$T_1 = 0$°C $= 273$ K  $\qquad\qquad T_2 = ?$ °C

$$T_2 = \frac{V_2T_1}{V_1} = \frac{(50.0 \text{ L})(273 \text{ K})}{25.0 \text{ L}} = 546 \text{ K} = 273\text{°C}$$

b.  $V_1 = 247$ mL  $\qquad\qquad V_2 = 255$ mL

$T_1 = 25$°C $= 298$ K  $\qquad\qquad T_2 = ?$°C

$$T_2 = \frac{V_2T_1}{V_1} = \frac{(255 \text{ mL})(298 \text{ K})}{247 \text{ mL}} = 308 \text{ K} = 35\text{°C}$$

c.  $V_1 = 1.00$ mL  $\qquad\qquad V_2 = ?$ mL

$T_1 = 2272$°C $= 2545$ K  $\qquad\qquad T_2 = 25$°C $= 298$ K

$$V_2 = \frac{V_1T_2}{T_1} = \frac{(1.00 \text{ mL})(298 \text{ K})}{2545 \text{ K}} = 0.117 \text{ mL}$$

## Chapter 13: Gases

34. a. $V_1 = 2.01 \times 10^2$ L $\qquad\qquad V_2 = 5.00$ L

   $T_1 = 1150°C = 1423$ K $\qquad T_2 = ?°C$

   $T_2 = \dfrac{V_2 T_1}{V_1} = \dfrac{(5.00 \text{ L})(1423 \text{ K})}{(201 \text{ L})} = 35.4$ K $= -238°C$

   b. $V_1 = 44.2$ mL $\qquad\qquad V_2 = ?$ mL

   $T_1 = 298$ K $\qquad\qquad T_2 = 0$

   $V_2 = \dfrac{V_1 T_2}{T_1} = \dfrac{(44.2 \text{ mL})(0 \text{ K})}{(298 \text{ K})} = 0$ mL (0 K is absolute zero)

   c. $V_1 = 44.2$ mL $\qquad\qquad V_2 = ?$ mL

   $T_1 = 298$ K $\qquad\qquad T_2 = 0°C = 273$ K

   $V_2 = \dfrac{V_1 T_2}{T_1} = \dfrac{(44.2 \text{ mL})(273 \text{ K})}{(298 \text{ K})} = 40.5$ mL

36. $V_2 = \dfrac{V_1 T_2}{T_1} = \dfrac{(125 \text{ mL})(250 \text{ K})}{(450 \text{ K})} = 69.4$ mL $= 69$ mL to two significant figures

38. $V_2 = \dfrac{V_1 T_2}{T_1}$

| Temp, °C | 90 | 80 | 70 | 60 | 50 | 40 | 30 | 20 |
|---|---|---|---|---|---|---|---|---|
| Volume, mL | 124 | 120. | 117 | 113 | 110 | 107 | 103 | 100 |

40. $V = an$; $V_1/n_1 = V_2/n_2$

42. $V = an$; $V_1/n_1 = V_2/n_2$

   Since 2.08 g of chlorine contains twice the number of moles of gas contained in the 1.04 g sample, the volume of the 2.08 g sample will be twice as large = 1744 ($1.74 \times 10^3$) mL

44. molar mass of Ar = 39.95 g

   $2.71 \text{ g Ar} \times \dfrac{1 \text{ mol}}{39.95 \text{ g}} = 0.0678$ mol Ar

   $4.21 \text{ L} \times \dfrac{1.29 \text{ mol}}{0.0678 \text{ mol}} = 80.1$ L

46. Real gases most closely approach ideal gas behavior under conditions of relatively high temperatures (0°C or higher) and relatively low pressures (1 atm or lower).

48. For an ideal gas, $PV = nRT$ is true under any conditions. Consider a particular sample of gas (so that $n$ remains constant) at a particular fixed pressure (so that $P$ remains constant also). Suppose that at temperature $T_1$ the volume of the gas sample is $V_1$. Then for this set of conditions, the ideal gas equation would be given by

   $PV_1 = nRT_1$.

Chapter 13: Gases

If we then change the temperature of the gas sample to a new temperature $T_2$, the volume of the gas sample changes to a new volume $V_2$. For this new set of conditions, the ideal gas equation would be given by

$PV_2 = nRT_2$.

If we make a ratio of these two expressions for the ideal gas equation for this gas sample, and cancel out terms that are constant for this situation ($P$, $n$, and $R$) we get

$$\frac{PV_1}{PV_2} = \frac{nRT_1}{nRT_2}$$

$$\frac{V_1}{V_2} = \frac{T_1}{T_2}$$

This can be rearranged to the familiar form of Charles's law

$$\frac{V_1}{T_1} = \frac{V_2}{T_2}$$

50. a. $P = 782$ mm Hg $= 1.03$ atm; $T = 27°C = 300$ K

$$V = \frac{nRT}{P} = \frac{(0.210 \text{ mol})(0.08206 \text{ L atm mol}^{-1}\text{ K}^{-1})(300 \text{ K})}{(1.03 \text{ atm})} = 5.02 \text{ L}$$

b. $V = 644$ mL $= 0.644$ L

$$P = \frac{nRT}{V} = \frac{(0.0921 \text{ mol})(0.08206 \text{ L atm mol}^{-1}\text{ K}^{-1})(303 \text{ K})}{(0.644 \text{ L})} = 3.56 \text{ atm}$$

$= 2.70 \times 10^3$ mm Hg

c. $P = 745$ mm $= 0.980$ atm

$$T = \frac{PV}{nR} = \frac{(0.980 \text{ atm})(11.2 \text{ L})}{(0.401 \text{ mol})(0.08206 \text{ L atm mol}^{-1}\text{ K}^{-1})} = 334 \text{ K}$$

52. Molar mass of $O_2 = 32.00$ g; 56.2 kg $= 5.62 \times 10^4$ g

$T = 21°C = 294$ K

$$n = 5.62 \times 10^4 \text{ g } O_2 \times \frac{1 \text{ mol } O_2}{32.00 \text{ g } O_2} = 1.76 \times 10^3 \text{ mol } O_2$$

$$P = \frac{nRT}{V} = \frac{(1.76 \times 10^3 \text{ mol})(0.08206 \text{ L atm mol}^{-1}\text{K}^{-1})(294 \text{ K})}{(125 \text{ L})} = 339 \text{ atm}$$

54. molar mass Ne $= 20.18$ g; 5.00 g $= 0.248$ mol

$$T = \frac{PV}{nR} = \frac{(1.10 \text{ atm})(7.00 \text{ L})}{(0.248 \text{ mol})(0.08206 \text{ L atm mol}^{-1}\text{K}^{-1})} = 379 \text{ K} = 106°C$$

Chapter 13: Gases

56. molar mass Ne = 20.18 g; 25°C = 298 K; 50°C = 323 K

$$1.25 \text{ g Ne} \times \frac{1 \text{ mol}}{20.18 \text{ g}} = 0.06194 \text{ mol}$$

$$P = \frac{nRT}{V} = \frac{(0.06194 \text{ mol})(0.08206 \text{ L atm mol}^{-1} \text{ K}^{-1})(298 \text{ K})}{(10.1 \text{ L})} = 0.150 \text{ atm}$$

$$P = \frac{nRT}{V} = \frac{(0.06194 \text{ mol})(0.08206 \text{ L atm mol}^{-1} \text{ K}^{-1})(323 \text{ K})}{(10.1 \text{ L})} = 0.163 \text{ atm}$$

58. molar mass $O_2$ = 32.00 g; 784 mm Hg = 1.032 atm

$$4.25 \text{ g } O_2 \times \frac{1 \text{ mol } O_2}{32.00 \text{ g } O_2} = 0.1328 \text{ mol}$$

$$T = \frac{PV}{nR} = \frac{(1.032 \text{ atm})(2.51 \text{ L})}{(0.1328 \text{ mol})(0.08206 \text{ L atm mol}^{-1} \text{ K}^{-1})} = 238 \text{ K} = -35°C$$

60. Molar masses: He, 4.003 g; Ar, 39.95 g

$$4.15 \text{ g He} \times \frac{1 \text{ mol He}}{4.003 \text{ g He}} = 1.037 \text{ mol He}$$

$$56.2 \text{ g Ar} \times \frac{1 \text{ mol Ar}}{39.95 \text{ g Ar}} = 1.407 \text{ mol Ar}$$

For He, $P = \dfrac{nRT}{V} = \dfrac{(1.037 \text{ mol})(0.08206 \text{ L atm mol}^{-1} \text{ K}^{-1})(298 \text{ K})}{(5.00 \text{ L})} = 5.07 \text{ atm}$

For Ar, $P = \dfrac{nRT}{V} = \dfrac{(1.407 \text{ mol})(0.08206 \text{ L atm mol}^{-1} \text{ K}^{-1})(303 \text{ K})}{(10.00 \text{ L})} = 3.50 \text{ atm}$

The helium is at a higher pressure than the argon.

62. molar mass Ar = 39.95 g; 29°C = 302 K; 42°C = 315 K

$$1.29 \text{ g Ar} \times \frac{1 \text{ mol Ar}}{39.95 \text{ g Ar}} = 0.03229 \text{ mol Ar}$$

$$P = \frac{nRT}{V} = \frac{(0.03229 \text{ mol})(0.08206 \text{ L atm mol}^{-1} \text{K}^{-1})(302 \text{ K})}{(2.41 \text{ L})} = 0.332 \text{ atm}$$

$$P = \frac{nRT}{V} = \frac{(0.03229 \text{ mol})(0.08206 \text{ L atm mol}^{-1} \text{K}^{-1})(315 \text{ K})}{(2.41 \text{ L})} = 0.346 \text{ atm}$$

64. Molar mass of $H_2O$ = 18.02 g; 2.0 mL = 0.0020 L; 225°C = 498 K

$$0.250 \text{ g } H_2O \times \frac{1 \text{ mol } H_2O}{18.02 \text{ g } H_2O} = 0.01387 \text{ mol } H_2O$$

Chapter 13: Gases

$$P = \frac{nRT}{V} = \frac{(0.01387 \text{ mol})(0.08206 \text{ L atm mol}^{-1}\text{ K}^{-1})(498 \text{ K})}{(0.0020 \text{ L})} = 283 \text{ atm} = 2.8 \times 10^2 \text{ atm}$$

66. As a gas is bubbled through water, the bubbles of gas become saturated with water vapor, thus forming a gaseous mixture. The total pressure in a sample of gas that has been collected by bubbling through water is made up of two components: the pressure of the gas of interest and the pressure of water vapor. The partial pressure of the gas of interest is then the total pressure of the sample minus the vapor pressure of water.

68. molar masses: Ne, 20.18 g; Ar, 39.95 g; 27°C = 300 K

    $$1.28 \text{ g Ne} \times \frac{1 \text{ mol Ne}}{20.18 \text{ g Ne}} = 0.06343 \text{ mol Ne}$$

    $$2.49 \text{ g Ar} \times \frac{1 \text{ mol Ar}}{39.95 \text{ g Ar}} = 0.06233 \text{ mol Ar}$$

    $$P_{neon} = \frac{n_{neon}RT}{V} = \frac{(0.06343 \text{ mol})(0.08206 \text{ L atm mol}^{-1}\text{ K}^{-1})(300 \text{ K})}{(9.87 \text{ L})} = 0.1582 \text{ atm}$$

    $$P_{argon} = \frac{n_{argon}RT}{V} = \frac{(0.06233 \text{ mol})(0.08206 \text{ L atm mol}^{-1}\text{ K}^{-1})(300 \text{ K})}{(9.87 \text{ L})} = 0.1555 \text{ atm}$$

    $P_{total}$ = 0.1582 atm + 0.1555 atm = 0.314 atm

70. 925 mm Hg = 1.217 atm; 26°C = 299 K; molar masses: Ne, 20.18 g; Ar, 39.95 g

    $$n = \frac{PV}{RT} = \frac{(1.217 \text{ atm})(3.00 \text{ L})}{(0.08206 \text{ L atm mol}^{-1}\text{ K}^{-1})(299 \text{ K})} = 0.1488 \text{ mol}$$

    The number of moles of an ideal gas required to fill a given-sized container to a particular pressure at a particular temperature does not depend on the specific identity of the gas. So 0.1488 mol of Ne gas or 0.1488 mol of Ar gas would give the same pressure in the same flask at the same temperature.

    $$\text{mass Ne} = 0.1488 \text{ mol Ne} \times \frac{20.18 \text{ g Ne}}{1 \text{ mol Ne}} = 3.00 \text{ g Ne}$$

    $$\text{mass Ar} = 0.1488 \text{ mol Ar} \times \frac{39.95 \text{ g Ar}}{1 \text{ mol Ar}} = 5.94 \text{ g Ar}$$

72. The total pressure of the gases inside the container is 1.00 atm. The total number of gas particles inside the container is 10. Let $x$ equal the pressure of each gas particle.

    1.00 atm = 10$x$
    $x$ = 0.100 atm
    There are 2 argon particles, 3 neon particles, and 5 helium particles present in the container.
    Therefore, $P_{Total} = P_{Ar} + P_{Ne} + P_{He} = 2x + 3x + 5x = 10x$.
    $P_{Ar}$ = 2(0.100 atm) = 0.20 atm (taking into account significant figures when these pressures are added)
    $P_{Ne}$ = 3(0.100 atm) = 0.30 atm (taking into account significant figures when these pressures are added)
    $P_{He}$ = 5(0.100 atm) = 0.50 atm (taking into account significant figures when these pressures are added)

Chapter 13: Gases

74. 1.032 atm = 784.3 mm Hg; molar mass of Zn = 65.38 g

$P_{hydrogen}$ = 784.3 mm Hg − 32 mm Hg = 752.3 mm Hg = 0.990 atm

$V$ = 240 mL = 0.240 L; $T$ = 30°C + 273 = 303 K

$$n_{hydrogen} = \frac{PV}{RT} = \frac{(0.990 \text{ atm})(0.240 \text{ L})}{(0.08206 \text{ L atm mol}^{-1} \text{ K}^{-1})(303 \text{ K})} = 0.00956 \text{ mol hydrogen}$$

$$0.00956 \text{ mol H}_2 \times \frac{1 \text{ mol Zn}}{1 \text{ mol H}_2} = 0.00956 \text{ mol of Zn must have reacted}$$

$$0.00956 \text{ mol Zn} \times \frac{65.38 \text{ g Zn}}{1 \text{ mol Zn}} = 0.625 \text{ g Zn must have reacted}$$

76. A theory is successful if it explains known experimental observations. Theories that have been successful in the past may not be successful in the future (for example, as technology evolves, more sophisticated experiments may be possible in the future).

78. pressure

80. no

82. If the temperature of a sample of gas is increased, the average kinetic energy of the particles of gas increases. This means that the speeds of the particles increase. If the particles have a higher speed, they will hit the walls of the container more frequently and with greater force, thereby increasing the pressure.

84. Standard Temperature and Pressure, STP = 0°C, 1 atm pressure. These conditions were chosen because they are easy to attain and reproduce *experimentally*. The barometric pressure within a laboratory is likely to be near 1 atm most days, and 0°C can be attained with a simple ice bath.

86. Molar mass of C = 12.01 g; 25°C = 298 K

$$1.25 \text{ g C} \times \frac{1 \text{ mol}}{12.01 \text{ g}} = 0.1041 \text{ mol C}$$

Since the balanced chemical equation shows a 1:1 stoichiometric relationship between C and $O_2$, then 0.1041 mol of $O_2$ will be needed

$$V = \frac{nRT}{P} = \frac{(0.1041 \text{ mol})(0.08206 \text{ L atm mol}^{-1} \text{ K}^{-1})(298 \text{ K})}{(1.02 \text{ atm})} = 2.50 \text{ L O}_2$$

88. Molar mass of Mg = 24.31 g; STP: 1.00 atm, 273 K

$$1.02 \text{ g Mg} \times \frac{1 \text{ mol}}{24.31 \text{ g}} = 0.0420 \text{ mol Mg}$$

As the coefficients for Mg and $Cl_2$ in the balanced equation are the same, for 0.0420 mol of Mg reacting we will need 0.0420 mol of $Cl_2$.

$$V = 0.0420 \text{ mol Cl}_2 \times \frac{22.4 \text{ L}}{1 \text{ mol}} = 0.940 \text{ L Cl}_2 \text{ at STP.}$$

Chapter 13: Gases

90. molar mass $CaC_2$ = 64.10 g; 25°C = 298 K

$$2.49 \text{ g CaC}_2 \times \frac{1 \text{ mol}}{64.10 \text{ g}} = 0.03885 \text{ mol CaC}_2$$

From the balanced chemical equation for the reaction, 0.03885 mol of $CaC_2$ reacting completely would generate 0.03885 mol of acetylene, $C_2H_2$

$$V = \frac{nRT}{P} = \frac{(0.03885 \text{ mol})(0.08206 \text{ L atm mol}^{-1} \text{ K}^{-1})(298 \text{ K})}{(1.01 \text{ atm})} = 0.941 \text{ L}$$

$$V = \frac{nRT}{P} = \frac{(0.03885 \text{ mol})(0.08206 \text{ L atm mol}^{-1} \text{ K}^{-1})(273 \text{ K})}{(1.00 \text{ atm})} = 0.870 \text{ L at STP}$$

92. Molar mass of $Mg_3N_2$ = 100.95 g; T = 24°C = 297 K; P = 752 mm Hg = 0.989 atm

$$10.3 \text{ g Mg}_3\text{N}_2 \times \frac{1 \text{ mol}}{100.95 \text{ g}} = 0.102 \text{ mol Mg}_3\text{N}_2$$

From the balanced chemical equation, the amount of $NH_3$ produced will be

$$0.102 \text{ mol Mg}_3\text{N}_2 \times \frac{2 \text{ mol NH}_3}{1 \text{ mol Mg}_3\text{N}_2} = 0.204 \text{ mol NH}_3$$

$$V = \frac{nRT}{P} = \frac{(0.204 \text{ mol})(0.08206 \text{ L atm mol}^{-1} \text{ K}^{-1})(297 \text{ K})}{(0.989 \text{ atm})} = 5.03 \text{ L}$$

This assumes that the ammonia was collected dry.

94. The balanced chemical equation is: $2X(s) + 3F_2(g) \rightarrow 2XF_3(s)$

To determine the moles of $F_2(g)$ present, use the Ideal Gas Law, since $F_2$ is in the gas phase.

$$n = \frac{PV}{RT} = \frac{(2.50 \text{ atm})(4.00 \text{ L})}{(0.08206 \text{ L atm mol}^{-1}\text{K}^{-1})(250.+273 \text{ K})} = 0.233 \text{ mol F}_2$$

Determine how many moles of X were used based on the amount of $F_2$ and the balanced equation.

$$0.233 \text{ mol F}_2 \times \frac{2 \text{ mol X}}{3 \text{ mol F}_2} = 0.155 \text{ mol X}$$

Use the given mass of X and moles of X to determine the molar mass.

$$\frac{9.15 \text{ g X}}{0.155 \text{ mol X}} = 58.9 \text{ g/mol}$$

The molar mass identifies the element as cobalt (Co).

96. Molar mass: $NH_3$, 17.034 g; 11°C + 273 = 284 K

The balanced equation is: $N_2(g) + 3H_2(g) \rightarrow 2NH_3(g)$

$$4.00 \text{ g NH}_3 \times \frac{1 \text{ mol NH}_3}{17.034 \text{ g NH}_3} = 0.235 \text{ mol NH}_3$$

Chapter 13: Gases

$$0.235 \text{ mol NH}_3 \times \frac{1 \text{ mol N}_2}{2 \text{ mol NH}_3} = 0.117 \text{ mol N}_2$$

$$0.235 \text{ mol NH}_3 \times \frac{3 \text{ mol H}_2}{2 \text{ mol NH}_3} = 0.352 \text{ mol H}_2$$

0.117 mol + 0.352 mol = 0.469 mol total reactant gases

$$V = \frac{nRT}{P} = \frac{(0.469 \text{ mol})(0.08206 \text{ L atm mol}^{-1}\text{K}^{-1})(284 \text{ K})}{(0.998 \text{ atm})} = 11.0 \text{ L}$$

98. a. $10N_2O(g) + C_3H_8(g) \rightarrow 10N_2(g) + 3CO_2(g) + 4H_2O(g)$

    b. Solve for moles of $N_2O$ and $C_3H_8$ using $PV = nRT$.

    $$n_{N_2O} = \frac{PV}{RT} = \frac{(1.4 \text{ atm})(1.0 \text{ L})}{(0.08206 \text{ L atm mol}^{-1}\text{K}^{-1})(450+273 \text{ K})} = 0.024 \text{ mol N}_2\text{O}$$

    $$n_{C_3H_8} = \frac{PV}{RT} = \frac{(1.2 \text{ atm})(1.0 \text{ L})}{(0.08206 \text{ L atm mol}^{-1}\text{K}^{-1})(450+273 \text{ K})} = 0.020 \text{ mol C}_3\text{H}_8$$

    Determine the limiting reactant.

    $$0.024 \text{ mol N}_2\text{O} \times \frac{3 \text{ mol CO}_2}{10 \text{ mol N}_2\text{O}} = 0.0071 \text{ mol CO}_2$$

    $$0.020 \text{ mol C}_3\text{H}_8 \times \frac{3 \text{ mol CO}_2}{1 \text{ mol C}_3\text{H}_8} = 0.061 \text{ mol CO}_2$$

    The maximum amount of $CO_2$ that can be produced is 0.0071 mol, thus $N_2O$ is the limiting reactant.

    Now solve for the pressure of $CO_2$ using $PV = nRT$. The total volume is 2.0 L since the valve was opened (1.0 L + 1.0 L = 2.0 L).

    $$P = \frac{nRT}{V} = \frac{(0.0071 \text{ mol})(0.08206 \text{ L atm mol}^{-1}\text{K}^{-1})(450+273\text{K})}{2.0 \text{ L}} = 0.21 \text{ atm CO}_2$$

    c. The partial pressure of $N_2O$ after the reaction is complete is 0 atm because it is completely consumed as the limiting reactant.

100. $2C_2H_2(g) + 5O_2(g) \rightarrow 2H_2O(g) + 4CO_2(g)$

    molar mass $C_2H_2$ = 26.04 g

    $$1.00 \text{ g C}_2\text{H}_2 \times \frac{1 \text{ mol}}{26.04 \text{ g}} = 0.0384 \text{ mol C}_2\text{H}_2$$

    From the balanced chemical equation, 2 × 0.0384 = 0.0768 mol of $CO_2$ will be produced.

    $$0.0768 \text{ mol CO}_2 \times \frac{22.4 \text{ L}}{1 \text{ mol}} = 1.72 \text{ L at STP}$$

102. 125 mL = 0.125 L

$$0.125 \text{ L} \times \frac{1 \text{ mol}}{22.4 \text{ L}} = 0.00558 \text{ mol } H_2$$

From the balanced chemical equation, one mole of zinc is required for each mole of hydrogen produced. Therefore, 0.00558 mol of Zn will be required.

$$0.00558 \text{ mol Zn} \times \frac{65.38 \text{ g Zn}}{1 \text{ mol}} = 0.365 \text{ g Zn}$$

104. (a)

106. In both cases, the gas particles will uniformly distribute throughout both flasks.

Case 1:

$$P_1V_1 = P_2V_2$$
$$P_1(2X) = P_2(3X)$$
$$\left(\frac{2}{3}\right)P_1 = P_2$$

Case 2:

$$P_1V_1 = P_2V_2$$
$$P_1(X) = P_2(2X)$$
$$\left(\frac{1}{2}\right)P_1 = P_2$$

Case 1: The drawing will have four particles in the left and two particles in the right.

Case 2: The drawing will have three particles in the left and three particles in the right.

108. First determine what volume the helium in the tank would have if it were at a pressure of 755 mm Hg (corresponding to the pressure the gas will have in the balloons).

8.40 atm = 6384 mm Hg

$$V_2 = (25.2 \text{ L}) \times \frac{6384 \text{ mm Hg}}{755 \text{ mm Hg}} = 213 \text{ L}$$

Allowing for the fact that 25.2 L of He will have to remain in the tank, this leaves 213 − 25.2 = 187.8 L of He for filling the balloons.

$$187.8 \text{ L He} \times \frac{1 \text{ balloon}}{1.50 \text{ L He}} = 125 \text{ balloons}$$

110. According to the balanced chemical equation, when 1 mol of $(NH_4)_2CO_3$ reacts, a total of 4 moles of gaseous substances is produced.

molar mass $(NH_4)_2CO_3$ = 96.09 g; 453 °C = 726 K

$$52.0 \text{ g} \times \frac{1 \text{ mol}}{96.09 \text{ g}} = 0.541 \text{ mol}$$

As 0.541 mol of $(NH_4)_2CO_3$ reacts, 4(0.541) = 2.16 mol of gaseous products result.

$$V = \frac{nRT}{P} = \frac{(2.16 \text{ mol})(0.08206 \text{ L atm mol}^{-1} \text{ K}^{-1})(726 \text{ K})}{(1.04 \text{ atm})} = 124 \text{ L}$$

Chapter 13: Gases

112. $CaCO_3(s) + 2H^+(aq) \rightarrow Ca^{2+}(aq) + H_2O(l) + CO_2(g)$

molar mass $CaCO_3 = 100.1$ g; $60°C + 273 = 333$ K

$10.0 \text{ g CaCO}_3 \times \dfrac{1 \text{ mol}}{100.1 \text{ g}} = 0.0999 \text{ mol CaCO}_3 = 0.0999 \text{ mol CO}_2$ also

$P_{\text{carbon dioxide}} = P_{\text{total}} - P_{\text{water vapor}}$

$P_{\text{carbon dioxide}} = 774 \text{ mm Hg} - 149.4 \text{ mm Hg} = 624.6 \text{ mm Hg} = 0.822$ atm

$V_{\text{wet}} = \dfrac{nRT}{P} = \dfrac{(0.0999 \text{ mol})(0.08206 \text{ L atm mol}^{-1} \text{ K}^{-1})(333 \text{ K})}{(0.822 \text{ atm})} = 3.32 \text{ L wet CO}_2$

$V_{\text{dry}} = 3.32 \text{ L} \times \dfrac{624.6 \text{ mm Hg}}{774 \text{ mm Hg}} = 2.68 \text{ L}$

114. (a); In order for the pressure in the balloon to double, the volume must be cut in half. However, the temperature in °C is also decreasing so the volume must decrease by more than half overall.

116. $n_1 = 0.214$ mol           $n_2 = 0.375$ mol

$V_1 = 652$ mL             $V_2 = ?$

$V_2 = \dfrac{V_1 n_2}{n_1} = \dfrac{(652 \text{ mL})(0.375 \text{ mol})}{0.214 \text{ mol}} = 1140 \text{ mL} = 1.14 \text{ L}$

118. a.
$PV = nRT$

$n = \dfrac{PV}{RT} = \dfrac{(1.00 \text{ atm})(2.50 \text{ L})}{(0.08206 \text{ L} \cdot \text{atm/mol} \cdot \text{K})(-48+273 \text{ K})}$

$n = 0.135$ mol

Molar Mass $= \dfrac{5.41 \text{ g}}{0.135 \text{ mol}} = 40.0$ g/mol

The gas is argon (Ar).

b. *Temperature of gas mixture*: $-48°C + 273 = 225$ K

*Total moles of gas mixture*: 0.448 mol

$10.0 \text{ g O}_2 \times \dfrac{1 \text{ mol O}_2}{32.00 \text{ g O}_2} = 0.3125 \text{ mol O}_2$

$0.313 \text{ mol} + 0.135 \text{ mol} = 0.448$ mol total

*Total pressure of gas mixture*: 1.00 atm

*Volume of balloon*: 8.27 L

$\dfrac{V_1}{n_1} = \dfrac{V_2}{n_2} \qquad \dfrac{2.50 \text{ L}}{0.135 \text{ mol}} = \dfrac{V_2}{0.448 \text{ mol}}$

$V_2 = 8.27$ L (wait until the end to round with significant figures)

Because the particles have kinetic energy and are in constant motion, adding more particles (moles of gas) will cause more collisions and thus a momentary increase in pressure on the balloon. The balloon will expand in volume to equalize with atmospheric pressure and keep pressure essentially constant.

c. *Temperature of gas mixture*: −48°C + 273 = 225 K

*Total moles of gas mixture*: 0.448 mol

$$10.0 \text{ g O}_2 \times \frac{1 \text{ mol O}_2}{32.00 \text{ g O}_2} = 0.3125 \text{ mol O}_2$$

0.313 mol + 0.135 mol = 0.448 mol total

*Total pressure of gas mixture*: 3.31 atm

$$\frac{P_1}{n_1} = \frac{P_2}{n_2} \quad \frac{1.00 \text{ atm}}{0.135 \text{ mol}} = \frac{P_2}{0.448 \text{ mol}}$$

$P_2$ = 3.31 atm (wait until the end to round with significant figures)

*Volume of rigid container*: 2.50 L

Because the particles have kinetic energy and are in constant motion, adding more particles (moles of gas) will cause more collisions and thus an increase in pressure on the container. Since the container is rigid and cannot expand, the volume remains constant while the pressure increases.

120. $1.05 \times 10^4$ mL; When the moles of gas and temperature are constant, pressure and volume are inversely related. Thus, if the pressure on a $2.10 \times 10^4$ mL sample of gas is doubled, the volume must be cut in half.

122. $P_1$ = 0.780 atm $\qquad P_2$ = ? atm

$V_1$ = 0.501 L $\qquad V_2$ = 0.794 L

$$P_2 = \frac{P_1 V_1}{V_2} = \frac{(0.780 \text{ atm})(0.501 \text{ L})}{0.794 \text{ L}} = 0.492 \text{ atm}$$

124. Molar Mass of Ar: 39.95 g; 25°C + 273 = 298 K

Since the question is only concerned with the partial pressure of Ar, the mass of He is not needed.

$$5.75 \text{ g Ar} \times \frac{1 \text{ mol Ar}}{39.95 \text{ g Ar}} = 0.144 \text{ mol Ar}$$

$$P = \frac{nRT}{V} = \frac{(0.144 \text{ mol})(0.08206 \text{ L atm mol}^{-1}\text{K}^{-1})(298 \text{ K})}{2.05 \text{ L}} = 1.72 \text{ atm Ar}$$

126. 126. $V_1$ = 0.475 L $\qquad V_2$ = ? L

$T_1$ = 27°C + 273 = 300. K $\qquad T_2$ = 82°C + 273 = 355 K

Chapter 13: Gases

$$\frac{V_1}{T_1} = \frac{V_2}{T_2} \quad \frac{0.475 \text{ L}}{300. \text{ K}} = \frac{V_2}{355 \text{ K}}$$

$$V_2 = 0.562 \text{ L}$$

128. Three changes you can make to double the volume are:

   1) *increase the temperature (double the temperature in the kelvin scale)*; If the temperature is increased, the gas particles have more kinetic energy and will hit the piston with more force (and more pressure). Therefore, the piston will move up until the pressure inside the container is the same as outside the container (causing the volume to increase).
   2) *add more moles of gas to the container (double the amount)*; By adding more moles of gas to the container, gas particles will hit the walls of the container more frequently (and thus exert more pressure). The piston will move up until the pressure inside the container is the same as outside the container (causing the volume to increase).
   3) *decrease the pressure outside the container (by half)*; By decreasing the pressure outside the container, the pressure inside becomes greater than the pressure outside. The gas particles inside will push the piston up until the pressure inside the container is the same as outside the container (causing the volume to increase).

130. a. $V = 21.2$ mL $= 0.0212$ L

   $$T = \frac{PV}{nR} = \frac{(1.034 \text{ atm})(0.0212 \text{ L})}{(0.00432 \text{ mol})(0.08206 \text{ L atm mol}^{-1} \text{ K}^{-1})} = 61.8 \text{ K}$$

   b. $V = 1.73$ mL $= 0.00173$ L

   $$P = \frac{nRT}{V} = \frac{(0.000115 \text{ mol})(0.08206 \text{ L atm mol}^{-1} \text{ K}^{-1})(182 \text{ K})}{(0.00173 \text{ L})} = 0.993 \text{ atm}$$

   c. $P = 1.23$ mm Hg $= 0.00162$ atm; $T = 152°\text{C} + 273 = 425$ K

   $$V = \frac{nRT}{P} = \frac{(0.773 \text{ mol})(0.08206 \text{ L atm mol}^{-1} \text{ K}^{-1})(425 \text{ K})}{(0.00162 \text{ atm})} = 1.66 \times 10^4 \text{ L}$$

132. Molar mass He: 4.003 g; Molar mass $H_2$: 2.016 g; 24°C + 273 = 297 K

   $$n = \frac{PV}{RT} = \frac{(2.70 \text{ atm})(100.0 \text{ L})}{(0.08206 \text{ L atm mol}^{-1}\text{K}^{-1})(297 \text{ K})} = 11.1 \text{ mol}$$

   $$11.1 \text{ mol He} \times \frac{4.003 \text{ g He}}{1 \text{ mol He}} = 44.3 \text{ g He}$$

   $$11.1 \text{ mol H}_2 \times \frac{2.016 \text{ g H}_2}{1 \text{ mol H}_2} = 22.3 \text{ g H}_2$$

134. Molar mass of Ne = 20.18 g; Molar mass of Ar = 39.95 g

   $$22 \text{ g Ne} \times \frac{1 \text{ mol Ne}}{20.18 \text{ g Ne}} = 1.1 \text{ mol Ne}$$

$$44 \text{ g Ar} \times \frac{1 \text{ mol Ar}}{39.95 \text{ g Ar}} = 1.1 \text{ mol Ar}$$

Since the same number of moles of Ar is present, it will also exert the same pressure as Ne, 2.0 atm. At constant temperature and volume, there is a direct relationship between the moles of gas and pressure exerted by the gas.

136.   a.   The can does not explode or crumple because the pressure inside the can and the pressure outside the can (from atmosphere pressure) continue to remain "balanced" or equalized. Even though more gaseous molecules are added to the can from boiling the water, they have enough kinetic energy to escape out the opening and remain equalized with the outside pressure.

   b.   The can crumples because as the can cools, the gaseous water molecules do not have as much kinetic energy and some condense to a liquid. Therefore, less gas molecules are present in the can and liquid water molecules do not take up as much space (volume). So essentially there are less moles of gas in the can colliding with the walls of the container and thus not as much pressure is exerted. Now the pressure inside the container is less than the pressure outside (from atmospheric pressure) and the outside pressure exerts more force and crushes the can.

   c.
$$PV = nRT$$
$$n = \frac{PV}{RT} = \frac{(1.00 \text{ atm})(4.00 \text{ L})}{(0.08206 \text{ L} \cdot \text{atm/mol} \cdot \text{K})(100.+273 \text{ K})}$$
$$n = 0.131 \text{ mol H}_2\text{O vapor}$$

   d.
$$PV = nRT$$
$$n = \frac{PV}{RT} = \frac{(1.00 \text{ atm})(1.50 \text{ L})}{(0.08206 \text{ L} \cdot \text{atm/mol} \cdot \text{K})(35+273 \text{ K})}$$
$$n = 0.0593 \text{ mol H}_2\text{O vapor remains}$$

   moles of condensed $H_2O$ = 0.131 mol − 0.0593 mol = 0.072 mol condensed $H_2O$

$$0.072 \text{ mol H}_2\text{O} \times \frac{18.016 \text{ g H}_2\text{O}}{1 \text{ mol H}_2\text{O}} = 1.3 \text{ g H}_2\text{O condensed}$$

138.   a.
$$n = \frac{PV}{RT} = \frac{(8.35 \text{ atm})(48.2 \text{ L})}{(0.08206 \text{ L} \cdot \text{atm/mol} \cdot \text{K})(25 + 273 \text{ K})} = 16.5 \text{ mol N}_2$$

$$16.5 \text{ mol N}_2 \times \frac{6.022 \times 10^{23} \text{ N}_2 \text{ molecules}}{1 \text{ mol N}_2} = 9.91 \times 10^{24} \text{ N}_2 \text{ molecules}$$

b.

$$n = \frac{PV}{RT} = \frac{(8.71 \text{ atm})(48.2 \text{ L} + 22.0 \text{ L})}{(0.08206 \text{ L} \cdot \text{atm/mol} \cdot \text{K})(25 + 273 \text{ K})} = 25.0 \text{ mol}$$

c.

$$n_{TOT} = 25.0 \text{ mol total} = n_{N_2} + n_{He} = 16.5 \text{ mol} + n_{He}$$

$$n_{He} = 8.5 \text{ mol He}$$

$$P = \frac{nRT}{V} = \frac{(8.5 \text{ mol})(0.08206 \text{ L} \cdot \text{atm/mol} \cdot \text{K})(298 \text{ K})}{22.0 \text{ L}} = 9.4 \text{ atm}$$

d.

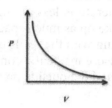

$n$ and $T$ are constant before and after the valve is connected. Thus, as the volume expands into two cylinders, the helium spreads throughout and the pressure decreases, showing an inverse relationship.

140. a. $CaC_2(s) + 2H_2O(l) \rightarrow C_2H_2(g) + Ca(OH)_2(aq)$

$2C_2H_2(g) + 5O_2(g) \rightarrow 4CO_2(g) + 2H_2O(g)$

b. molar mass of $CaC_2$ = 64.10 g; molar mass of $H_2O$ = 18.016 g

$$100.0 \text{ g CaC}_2 \times \frac{1 \text{ mol CaC}_2}{64.10 \text{ g CaC}_2} = 1.560 \text{ mol CaC}_2$$

$$50.0 \text{ g H}_2\text{O} \times \frac{1 \text{ mol H}_2\text{O}}{18.016 \text{ g H}_2\text{O}} = 2.78 \text{ mol H}_2\text{O}$$

$$1.560 \text{ mol CaC}_2 \times \frac{2 \text{ mol H}_2\text{O}}{1 \text{ mol CaC}_2} = 3.120 \text{ mol H}_2\text{O required to react}$$

Since only 2.78 mol $H_2O$ are available and 3.120 mol are needed to react with all of the $CaC_2$ present, $H_2O$ is the limiting reactant.

c. $H_2O$ is the limiting reactant and thus determines how much $C_2H_2$ is produced.

$$2.78 \text{ mol H}_2\text{O} \times \frac{1 \text{ mol C}_2\text{H}_2}{2 \text{ mol H}_2\text{O}} = 1.39 \text{ mol C}_2\text{H}_2$$

$$1.39 \text{ mol C}_2\text{H}_2 \times \frac{4 \text{ mol CO}_2}{2 \text{ mol C}_2\text{H}_2} = 2.78 \text{ mol CO}_2$$

Now determine how many liters of carbon dioxide gas were produced in the balloon at a pressure of 1.00 atm and 25°C:

$$V = \frac{nRT}{P}$$

$$V = \frac{(2.78 \text{ mol})(0.08206 \text{ L} \cdot \text{atm/mol} \cdot \text{K})(25 + 273 \text{ K})}{(1.00 \text{ atm})}$$

$$V = 67.9 \text{ L CO}_2$$

142. $2Cu_2S(s) + 3O_2(g) \rightarrow 2Cu_2O(s) + 2SO_2(g)$

molar mass $Cu_2S$ = 159.2 g; 27.5°C + 273 = 301 K

$$25 \text{ g Cu}_2\text{S} \times \frac{1 \text{ mol Cu}_2\text{S}}{159.2 \text{ g Cu}_2\text{S}} = 0.1570 \text{ mol Cu}_2\text{S}$$

$$0.1570 \text{ mol Cu}_2\text{S} \times \frac{3 \text{ mol O}_2}{2 \text{ mol Cu}_2\text{S}} = 0.2355 \text{ mol O}_2$$

$$V_{\text{oxygen}} = \frac{nRT}{P} = \frac{(0.2355 \text{ mol})(0.08206 \text{ L atm mol}^{-1} \text{ K}^{-1})(301 \text{ K})}{(0.998 \text{ atm})} = 5.8 \text{ L O}_2$$

$$0.1570 \text{ mol Cu}_2\text{S} \times \frac{2 \text{ mol SO}_2}{2 \text{ mol Cu}_2\text{S}} = 0.1570 \text{ mol SO}_2$$

$$V_{\text{sulfur dioxide}} = \frac{nRT}{P} = \frac{(0.1570 \text{ mol})(0.08206 \text{ L atm mol}^{-1} \text{ K}^{-1})(301 \text{ K})}{(0.998 \text{ atm})} = 3.9 \text{ L SO}_2$$

144. One mole of any ideal gas occupies 22.4 L at STP.

$$35 \text{ mol N}_2 \times \frac{22.4 \text{ L}}{1 \text{ mol}} = 7.8 \times 10^2 \text{ L}$$

146. a. Assuming the temperature is the same on both sides, the initial pressure of helium is greater because there are more gas particles colliding with the walls of its container (within the same volume).

$$\frac{P_{\text{He}}}{P_{\text{Ne}}} = \frac{n_{\text{He}}}{n_{\text{Ne}}} = \frac{6}{4} = 1.5 \text{ times greater}$$

b. When the valve is opened, the gases disperse uniformly throughout the entire apparatus (i.e., two Ne and three He in each vessel).

c. From the diagram, when the volume doubles the pressure is halved so

$$P_{f(\text{Ne})} = \frac{1}{2} P_{o(\text{Ne})}$$

$$P_{f(\text{He})} = \frac{1}{2} P_{o(\text{He})}$$

Chapter 13: Gases

d. The original pressure of helium is 1.5 times the original pressure of neon.

$P_{o(He)} = 1.5\, P_{o(Ne)}$

The final pressure is

$$P_f = \frac{1}{2}P_{o(Ne)} + \frac{1}{2}P_{o(He)} = \frac{1}{2}P_{o(Ne)} + \frac{1.5}{2}P_{o(Ne)}$$

$$= \frac{2.5}{2}P_{o(Ne)} = 1.25 P_{o(Ne)}$$

148. The solution is only 50% $H_2O_2$. Therefore 125 g solution = 62.5 g $H_2O_2$

molar mass of $H_2O_2$ = 34.02 g; $T = 27°C = 300$ K; $P = 764$ mm Hg = 1.01 atm

$$62.5\text{ g } H_2O_2 \times \frac{1\text{ mol}}{34.02\text{ g}} = 1.84\text{ mol } H_2O_2$$

$$1.84\text{ mol } H_2O_2 \times \frac{1\text{ mol } O_2}{2\text{ mol } H_2O_2} = 0.920\text{ mol } O_2$$

$$V = \frac{nRT}{P} = \frac{(0.920\text{ mol})(0.08206\text{ L atm mol}^{-1}\text{ K}^{-1})(300\text{ K})}{(1.01\text{ atm})} = 22.4\text{ L}$$

150. $V = .04 \times 500$ mL = 20 mL $CO_2$ = 0.02 L

molar mass of $CO_2$ = 44.01 g; $T = 25°C = 298$ K; $P = 1.00$ atm

$$n = \frac{PV}{RT} = \frac{(1.00\text{ atm})(0.02\text{ L})}{(0.08206\text{ L atm mol}^{-1}\text{ K}^{-1})(298\text{ K})} = 8 \times 10^{-4}\text{ mol } CO_2$$

$$8 \times 10^{-4}\text{ mol } CO_2 \times \frac{44.01\text{ g } CO_2}{1\text{ mol } CO_2} = 0.04\text{ g } CO_2$$

152. (b), (c), (d)

molar mass of $N_2$ = 28.02 g; 28 g = 1.0 mol

molar mass of $O_2$ = 32.00 g; 28 g = 0.88 mol; 32 g = 1.0 mol

Double the moles of gas would need to be present in order to double the pressure. Adding 28 g of $O_2$ is not double the moles, as in (a), but adding 32 g of $O_2$ is double the moles, as in (d). ($n_1/P_1 = n_2/P_2$)

$T = -73°C = 200.$ K; $T = 127°C = 400.$ K; $T = 30°C = 303$ K; $T = 60°C = 333$ K

Doubling the temperature (in Kelvin) will double the pressure, as in (b), but not in (e). ($P_1/T_1 = P_2/T_2$)

Cutting the volume by half will double the pressure, as in (c). ($P_1V_1 = P_2V_2$)

154. $V_1 = 855$ L $\qquad\qquad$ $V_2 = ?$

$P_1 = 730$ torr $\qquad\qquad$ $P_2 = 605$ torr

$T_1 = 25°C = 298$ K $\qquad\qquad$ $T_2 = 15°C = 288$ K

Chapter 13: Gases

$$V_2 = \frac{P_1V_1T_2}{T_1P_2} = \frac{(730 \text{ torr})(855 \text{ L})(288 \text{ K})}{(298 \text{ K})(605 \text{ torr})} = 997 \text{ L}$$

$\Delta V = 997 \text{ L} - 855 \text{ L} = 142 \text{ L}$

156. $Xe + 2F_2 \rightarrow XeF_4$

    molar mass of $XeF_4$ = 207.3 g; $T = 400°C = 673$ K; $V = 20.0$ L;

    $P_{Xenon} = 0.859$ atm; $P_{Fluorine} = 1.37$ atm

    $$n_{Xenon} = \frac{PV}{RT} = \frac{(0.859 \text{ atm})(20.0 \text{ L})}{(0.08206 \text{ L atm mol}^{-1} \text{ K}^{-1})(673 \text{ K})} = 0.311 \text{ mol Xe}$$

    $$n_{Fluorine} = \frac{PV}{RT} = \frac{(1.37 \text{ atm})(20.0 \text{ L})}{(0.08206 \text{ L atm mol}^{-1} \text{ K}^{-1})(673 \text{ K})} = 0.496 \text{ mol F}_2$$

    $F_2$ is the limiting reactant (only 0.248 mol Xe needed).

    $0.496 \text{ mol F}_2 \times \dfrac{1 \text{ mol XeF}_4}{2 \text{ mol F}_2} \times \dfrac{207.3 \text{ g XeF}_4}{1 \text{ mol XeF}_4} = 51.4 \text{ g XeF}_4$

158. (*a*), (*c*), and (*d*); Only doubling the temperature in Kelvin will double the volume, thus (*b*) is not true.

# CHAPTER 14

# Liquids and Solids

2. Liquids and solids are *less* compressible than are gases.

4. Since it requires so much more energy to vaporize water than to melt ice, this suggests that the gaseous state is significantly different from the liquid state, but that the liquid and solid state are relatively similar.

6. See Figure 14.2.

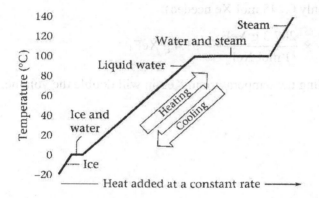

8. When a solid is heated, the molecules begin to vibrate/move more quickly. When enough energy has been added to overcome the intermolecular forces that hold the molecules in a crystal lattice, the solid melts. As the liquid is heated, the molecules begin to move more quickly and more randomly. When enough energy has been added, molecules having sufficient kinetic energy will begin to escape from the surface of the liquid. Once the pressure of vapor coming from the liquid is equal to the pressure above the liquid, the liquid boils. Only intermolecular forces need to be overcome in this process: no chemical bonds are broken.

10. Intramolecular; intermolecular

12. molar heat of fusion

14.  a. More energy is required to separate the atoms of the liquid into the freely-moving and widely-separated atoms of the vapor/gas.

 b. $1.00 \text{ g Al} \times \dfrac{1 \text{ mol Al}}{26.98 \text{ g Al}} \times \dfrac{293.4 \text{ kJ}}{1 \text{ mol Al}} = 10.9 \text{ kJ}$

 c. $5.00 \text{ g Al} \times \dfrac{1 \text{ mol Al}}{26.98 \text{ g Al}} \times \dfrac{-10.79 \text{ kJ}}{1 \text{ mol Al}} = -2.00 \text{ kJ}$ (heat is evolved)

 d. $0.105 \text{ mol Al} \times \dfrac{10.79 \text{ kJ}}{1 \text{ mol Al}} = 1.13 \text{ kJ}$

Chapter 14: Liquids and Solids

16. molar mass Ag = 107.9 g

melt: $12.5 \text{ g Ag} \times \dfrac{1 \text{ mol Ag}}{107.9 \text{ g Ag}} \times \dfrac{11.3 \text{ kJ}}{1 \text{ mol Al}} = 1.31 \text{ kJ}$

condense: $4.59 \text{ g Ag} \times \dfrac{1 \text{ mol Ag}}{107.9 \text{ g Ag}} \times \dfrac{-250. \text{ kJ}}{1 \text{ mol Ag}} = -10.6 \text{ kJ}$ (heat is evolved)

18. The *molar* heat of fusion of aluminum is the heat required to melt 1 mol.

$\dfrac{113 \text{ J}}{1.00 \text{ g Na}} \times \dfrac{22.99 \text{ g Na}}{1 \text{ mol Na}} = 2598 \text{ J/mol} = 2.60 \text{ kJ/mol}$

20. stronger

22. Water molecules are able to form strong *hydrogen bonds* with each other. These bonds are an especially strong form of dipole-dipole forces and are only possible when hydrogen atoms are bonded to the most electronegative elements (N, O, and F). The particularly strong intermolecular forces in $H_2O$ require much higher temperatures (higher energies) to be overcome in order to permit the liquid to boil. We take the fact that water has a much higher boiling point than the other hydrogen compounds of the Group 6 elements as proof that a special force is at play in water (hydrogen bonding).

24. London dispersion forces

26.  a. London dispersion forces (noble gas atoms)

b. hydrogen bonding (H bonded to N); London dispersion forces

c. London dispersion forces (nonpolar molecules)

d. dipole-dipole forces (polar molecules); London dispersion forces

28. An increase in the heat of fusion is observed for an increase in the size of the halogen atom involved (the electron cloud of a larger atom is more easily polarized by an approaching dipole, thus giving larger London dispersion forces).

30. For a homogeneous mixture to be able to form at all, the forces between molecules of the two substances being mixed must be at least *comparable in magnitude* to the intermolecular forces within each *separate* substance. Apparently in the case of a water-ethanol mixture, the forces that exist when water and ethanol are mixed are stronger than water-water or ethanol-ethanol forces in the separate substances. This allows ethanol and water molecules to approach each other more closely in the mixture than either substance's molecules could approach a like molecule in separate substances. There is strong hydrogen bonding in both ethanol and water.

32. Vapor pressure is the pressure of vapor present *at equilibrium* above a liquid in a sealed container at a particular temperature. When a liquid is placed in a closed container, molecules of the liquid evaporate freely into the empty space above the liquid. As the number of molecules present in the vapor state increases with time, vapor molecules begin to rejoin the liquid state (condense). Eventually a dynamic equilibrium is reached between evaporation and condensation in which the net number of molecules present in the vapor phase becomes *constant* with time. Since vapor pressure increases with temperature, the vapor pressure of the solvent is higher on a warm day.

## Chapter 14: Liquids and Solids

34. A liquid is injected at the bottom of the column of mercury and rises to the surface of the mercury, where the liquid evaporates into the vacuum above the mercury column. As the liquid evaporates, the pressure of vapor increases in the space above the mercury, and presses down on the mercury. The level of mercury, therefore, drops, and the amount by which the mercury level drops (in mm Hg) is equivalent to the vapor pressure of the liquid.

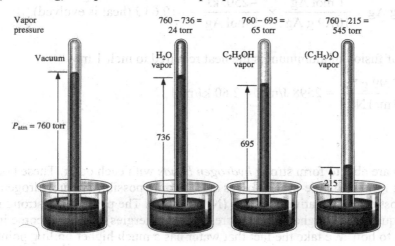

In the picture, the left tube represents a barometer—a tube of mercury inverted into a dish of mercury, with a vacuum above the mercury column: the height of the mercury column represents the pressure of the atmosphere. In the remaining 3 tubes, liquids of different volatilities are admitted to the bottom of the tube of mercury: they rise through the mercury and evaporate into the vacuum above the column of mercury. As the pressure of vapor builds up above the mercury column, the height of the mercury in the tube drops. Note that diethyl ether, $(C_2H_5)_2O$, shows the highest vapor pressure because it is the most volatile of the three liquids.

36. a. $H_2S$: $H_2O$ exhibits hydrogen bonding, and $H_2S$ does not. In general, substances that exhibit weaker intermolecular forces are more volatile.

    b. $CH_3OH$: $H_2O$ exhibits stronger hydrogen bonding than $CH_3OH$ because there are two locations where hydrogen bonding is possible on water.

    c. $CH_3OH$: Both are capable of hydrogen bonding, but generally lighter molecules are more volatile than heavier molecules.

38. Both substances have the same molar mass. However, ethyl alcohol contains a hydrogen atom directly bonded to an oxygen atom. Therefore, hydrogen bonding can exist in ethyl alcohol, whereas only weak dipole-dipole forces can exist in dimethyl ether. Dimethyl ether is more volatile; ethyl alcohol has a higher boiling point.

40. *Ionic* solids have as their fundamental particles positive and negative *ions*; a simple example is sodium chloride, in which $Na^+$ and $Cl^-$ ions are held together by strong electrostatic forces.

    *Molecular* solids have molecules as their fundamental particles, with the molecules being held together in the crystal by dipole-dipole forces, hydrogen bonding forces, or London dispersion forces (depending on the identity of the substance); simple examples of molecular solids include ice ($H_2O$) and ordinary table sugar (sucrose).

    *Atomic* solids have simple atoms as their fundamental particles, with the atoms being held together in the crystal either by covalent bonding (as in graphite or diamond) or by metallic bonding (as in copper or other metals).

42. The interparticle forces in ionic solids (the ionic bond) are much stronger than the interparticle forces in molecular solids (dipole-dipole forces, London forces, etc.). The difference in intermolecular forces is most clearly shown in the great differences in melting points and boiling points between ionic and molecular solids. For example, table salt and ordinary sugar are both crystalline solids that appear very similar. Yet sugar can be melted easily in a saucepan during the making of candy, whereas even the full heat of a stove will not melt salt.

44. Strong electrostatic forces exist between oppositely charged ions in ionic solids.

46. In liquid hydrogen, the only intermolecular forces are weak London dispersion forces. In ethyl alcohol and water we have hydrogen bonding possible, but the hydrogen bonding forces are weaker in ethyl alcohol because of the influence of the remainder of the molecule. In sucrose, we also have hydrogen bonding possible, but now at several places in the molecule, leading to stronger forces. In calcium chloride, we have an ionic crystal lattice with even stronger forces between the particles.

48. Although ions exist in the solid, liquid, or dissolved states, in the solid state the ions are rigidly held in place in the crystal lattice and cannot *move* so as to conduct an electrical current.

50. Nitinol is an alloy of nickel and titanium. When nickel and titanium are heated to a sufficiently high temperature during the production of Nitinol, the atoms arrange themselves in a compact and regular pattern of the atoms.

52. j

54. f

56. d

58. a

60. l

62. Diethyl ether has the larger vapor pressure. No hydrogen bonding is possible because the O atom does not have a hydrogen atom attached. Hydrogen bonding can occur *only* when a hydrogen atom is *directly* attached to a strongly electronegative atom (such as N, O, or F). Hydrogen bonding *is* possible in 1-butanol (1-butanol contains an ÷OH group).

64. None of the substances listed exhibit hydrogen bonding interactions.

   $CCl_2H_2$: dipole-dipole forces; London dispersion forces

   $BeF_2$: London dispersion forces

   $NO_3^-$: London dispersion forces

   HCN: dipole-dipole forces; London dispersion forces

66. *Steel* is a general term applied to alloys consisting primarily of iron, but with small amounts of other substances added. Whereas pure iron itself is relatively soft, malleable, and ductile, steels are typically much stronger and harder and much less subject to damage.

## Chapter 14: Liquids and Solids

68. Water is the solvent in which cellular processes take place in living creatures. Water in the oceans moderates the Earth's temperature. Water is used in industry as a cooling agent. Water serves as a means of transportation on the Earth's oceans. The liquid range is 0°C to 100°C at 1 atm pressure.

70. At higher altitudes, the boiling points of liquids, such as water, are lower because there is a lower atmospheric pressure above the liquid. The temperature at which food cooks is determined by the temperature to which the water in the food can be heated before it escapes as steam. Thus, food cooks at a lower temperature at high elevations where the boiling point of water is lowered.

72. Heat of fusion (melt); heat of vaporization (boil).

    The heat of vaporization is always larger, because virtually all of the intermolecular forces must be overcome to form a gas. In a liquid, considerable intermolecular forces remain. Thus going from a solid to liquid requires less energy than going from the liquid to the gas.

74. Dipole-dipole interactions are typically about 1% as strong as a covalent bond. Dipole-dipole interactions represent electrostatic attractions between portions of molecules that carry only a *partial* positive or negative charge. Such forces require the molecules that are interacting to come *near* enough to each other.

76. London dispersion forces are relatively weak forces that arise among noble gas atoms and in nonpolar molecules. London forces are due to *instantaneous dipoles* that develop when one atom (or molecule) momentarily distorts the electron cloud of another atom (or molecule). London forces are typically weaker than either permanent dipole-dipole forces or covalent bonds.

78. a. London dispersion forces (nonpolar atoms)
    b. hydrogen bonding (H attached to O); London dispersion forces
    c. dipole-dipole (polar molecules); London dispersion forces
    d. London dispersion forces (nonpolar molecules)

80. a.

| | $CH_3OH$ | $CH_4$ | $H_2O$ | $C_2H_6$ |
|---|---|---|---|---|
| Lewis Structure | H–C(H)(H)–O–H | H–C(H)(H)–H | H–O–H | H–C(H)(H)–C(H)(H)–H |
| Polar? | Polar | Nonpolar | Polar | Nonpolar |

b. $H_2O$ and $CH_3OH$ are liquids at room temperature. $CH_4$ and $C_2H_6$ are gases at room temperature. $H_2O$ and $CH_3OH$ (the liquids) contain dipole-dipole interactions (specifically hydrogen bonding) and London dispersion forces. $CH_4$ and $C_2H_6$ (the gases) contain only London dispersion forces. Dipole-dipole interactions (or hydrogen bonding) are generally stronger than London dispersion forces for molecules that are about the same size.

c. (lowest bp) $CH_4 < C_2H_6 < CH_3OH < H_2O$ (highest bp)

The liquids have a higher boiling point than the gases, so gases < liquids. $CH_4 < C_2H_6$ because $C_2H_6$ has more electrons (or higher molar mass) and thus stronger London dispersion forces. $CH_3OH < H_2O$ because water has two O–H bonds and methanol has one O–H bond (could also talk about the nonpolarity/larger steric hindrance or crowding of $CH_3$ versus the other OH bond on water).

82. In a crystal of ice, strong *hydrogen bonding* forces are present, whereas in the crystal of a nonpolar substance like oxygen, only the much weaker *London* forces exist.

84. Ice floats on liquid water; water expands when it is frozen

86. Although they are at the same *temperature*, steam at 100°C contains a larger amount of *energy* than hot water, equal to the heat of vaporization of water.

88. a. The following species are polar: $SeCl_2$, ICl, $NO_2^-$

Based on their Lewis structures, $SeCl_2$, ICl, $NO_2^-$ are polar and $BH_3$ is nonpolar.

b. None of the species exhibit hydrogen bonding interactions.

90. Evaporation and condensation are opposite processes. Evaporation is an endothermic process, condensation is an exothermic process. Evaporation requires an input of energy to provide the increased kinetic energy possessed by the molecules when they are in the gaseous state. Evaporation occurs when the molecules in a liquid are moving fast enough and randomly enough that molecules are able to escape from the surface of the liquid and become a gas.

92. Diamonds are made of only one element (carbon). The very strong covalent bonds among the carbon atoms in diamond lead to a giant molecule, and these types of substances are referred to as network solids.

94. (a), (d), and (e) are true; Molecules that only exhibit London dispersion forces can exist in any state of matter at room temperature depending on the molar mass of the molecule. $H_2O$ exhibits stronger hydrogen bonding interactions because it has a greater charge separation within the molecule (greater differences in electronegativities).

96. $CH_3Cl$, $CH_3CH_2Cl$, $CH_3CH_2CH_2Cl$, $CH_3CH_2CH_2CH_2Cl$

The larger the molar mass of the molecule, the more London dispersion forces present, which leads to a higher boiling point.

Chapter 14: Liquids and Solids

98. (b), (c), and (d) are true; LiF has stronger interactions (ionic) versus $H_2S$ (dipole-dipole) thus it has a lower vapor pressure. Similarly, MgO has stronger interactions (ionic) versus $CH_3CH_2OH$ (hydrogen bonding) thus it also has a lower vapor pressure. HF exhibits stronger interactions (hydrogen bonding) versus HBr (dipole-dipole) thus it has a lower vapor pressure. $Cl_2$ exhibits stronger interactions (London dispersion with a higher molar mass) versus Ar (London dispersion with a lower molar mass) thus it has a higher boiling point. Water is polar, thus HCl is more soluble in water than $CCl_4$ because HCl is also polar (whereas $CCl_4$ is nonpolar).

# CHAPTER 15

# Solutions

2. A heterogeneous mixture does not have a uniform composition: the composition varies in different places within the mixture. Examples of non–homogeneous mixtures include salad dressing (mixture of oil, vinegar, water, herbs, and spices) and granite (combination of minerals).

4. solvent; solutes

6. "Like dissolves like." The hydrocarbons in oil have intermolecular forces that are very different from those in water, and so the oil spreads out rather than dissolving in the water.

8. Carbon dioxide is somewhat soluble in water, especially if pressurized (otherwise, the soda you may be drinking while studying Chemistry would be "flat"). Carbon dioxide's solubility in water is approximately 1.5 g/L at 25°C under a pressure of approximately 1 atm. Although the carbon dioxide molecule overall is non-polar, that is because the two individual C–O bond dipoles cancel each other due to the linearity of the molecule. However, these bond dipoles are able to interact with water, making $CO_2$ more soluble in water than non-polar molecules such as $O_2$ or $N_2$ that do not possess individual bond dipoles.

10. unsaturated

12. large

14. 100.0

16. 6.11 mg = 0.00611 g

   a. $\dfrac{0.00611 \text{ g CaCl}_2}{(0.00611 \text{ g CaCl}_2 + 5.25 \text{ g water})} \times 100 = 0.116\% \text{ CaCl}_2$

   b. $\dfrac{0.00611 \text{ g CaCl}_2}{(0.00611 \text{ g CaCl}_2 + 52.5 \text{ g water})} \times 100 = 0.0116\% \text{ CaCl}_2$

   c. $\dfrac{6.11 \text{ g CaCl}_2}{(6.11 \text{ g CaCl}_2 + 52.5 \text{ g water})} \times 100 = 10.4\% \text{ CaCl}_2$

   d. $\dfrac{6.11 \text{ kg CaCl}_2}{(6.11 \text{ kg CaCl}_2 + 52.5 \text{ kg water})} \times 100 = 10.4\% \text{ CaCl}_2$

18. a. $525 \text{ g solution} \times \dfrac{3.91 \text{ g FeCl}_3}{100. \text{ g solution}} = 20.5 \text{ g FeCl}_3$

   525 g solution – 20.5 g $FeCl_3$ = 504.5 g water (505 g water)

Chapter 15: Solutions

b. $225 \text{ g solution} \times \dfrac{11.9 \text{ g sucrose}}{100. \text{ g solution}} = 26.8 \text{ g sucrose}$

225 g solution − 26.8 g sucrose = 198.2 g water (198 g water)

c. $1.45 \text{ kg} = 1.45 \times 10^3 \text{ g}$

$1.45 \times 10^3 \text{ g solution} \times \dfrac{12.5 \text{ g NaCl}}{100. \text{ g solution}} = 181.3 \text{ g NaCl (181 g NaCl)}$

$1.45 \times 10^3$ g solution − 181.3 g NaCl = 1268.7 g water ($1.27 \times 10^3$ g water)

d. $635 \text{ g solution} \times \dfrac{15.1 \text{ g KNO}_3}{100. \text{ g solution}} = 95.9 \text{ g KNO}_3$

635 g solution − 95.9 g $KNO_3$ = 539.1 g water (539 g water)

20. Percent means "per hundred". So the percentages in Question 19 represent the number of grams of the particular element present in 100. g of the alloy. Since 1.00 kg represents ten times the mass of 100. g, we would need 957 g Fe, 26.9 g C, and 16.5 g Cr to prepare 1.00 kg of the alloy.

22. $\dfrac{67.1 \text{ g CaCl}_2}{(67.1 \text{ g CaCl}_2 + 275 \text{ g water})} \times 100 = 19.6\% \text{ CaCl}_2$

24. To say that the solution is 6.25% KBr by mass, means that 100. g of the solution will contain 6.25 g KBr.

$125 \text{ g solution} \times \dfrac{6.25 \text{ g KBr}}{100. \text{ g solution}} = 7.81 \text{ g KBr}$

26. molar mass $O_2$ = 32.00 g

$1.00 \text{ g O}_2 \times \dfrac{1 \text{ mol}}{32.00 \text{ g}} = 0.03125 \text{ mol O}$

from the balanced chemical equation, it will take 2(0.03125) = 0.0625 mol $H_2O_2$ to produce this quantity of oxygen.

molar mass $H_2O_2$ = 34.02 g

$0.0625 \text{ mol H}_2\text{O}_2 \times \dfrac{34.02 \text{ g H}_2\text{O}_2}{1 \text{ mol H}_2\text{O}_2} = 2.13 \text{ g H}_2\text{O}_2$

$2.13 \text{ g H}_2\text{O}_2 \times \dfrac{100. \text{ g solution}}{3 \text{ g H}_2\text{O}_2} = \text{approximately 71 g}$

28. $1000 \text{ g} \times \dfrac{0.95 \text{ g stablizer}}{100. \text{ g}} = 9.5 \text{ g}$

30. 0.25 mol; 0.75 mol      [$AlCl_3(aq) \rightarrow Al^{3+}(aq) + 3Cl^-(aq)$]

Chapter 15: Solutions

32. The molarity represents the number of moles of solute per liter of solution: choice b is the only scenario that fulfills this definition.

34. Molarity = $\dfrac{\text{moles of solute}}{\text{liters of solution}}$

   a. 225 mL = 0.225 L

   $M = \dfrac{0.754 \text{ mol KNO}_3}{0.225 \text{ L}} = 3.35\ M$

   b. 10.2 mL = 0.0102 L

   $M = \dfrac{0.0105 \text{ mol CaCl}_2}{0.0102 \text{ L}} = 1.03\ M$

   c. $M = \dfrac{3.15 \text{ mol NaCl}}{5.00 \text{ L}} = 0.630\ M$

   d. 100. mL = 0.100 L

   $M = \dfrac{0.499 \text{ mol NaBr}}{0.100 \text{ L}} = 4.99\ M$

36. Molarity = $\dfrac{\text{moles of solute}}{\text{liters of solution}}$ ; molar mass of $CaCl_2$ = 110.98 g; 125 mL = 0.125 L

   a. $5.59 \text{ g CaCl}_2 \times \dfrac{1 \text{ mol CaCl}_2}{110.98 \text{ g CaCl}_2} = 0.05037 \text{ mol CaCl}_2$

   $M = \dfrac{0.05037 \text{ mol}}{0.125 \text{ L}} = 0.403\ M$

   b. $2.34 \text{ g CaCl}_2 \times \dfrac{1 \text{ mol CaCl}_2}{110.98 \text{ g CaCl}_2} = 0.02108 \text{ mol CaCl}_2$

   $M = \dfrac{0.02108 \text{ mol}}{0.125 \text{ L}} = 0.169\ M$

   c. $8.73 \text{ g CaCl}_2 \times \dfrac{1 \text{ mol CaCl}_2}{110.98 \text{ g CaCl}_2} = 0.07866 \text{ mol CaCl}_2$

   $M = \dfrac{0.07866 \text{ mol}}{0.125 \text{ L}} = 0.629\ M$

   d. $11.5 \text{ g CaCl}_2 \times \dfrac{1 \text{ mol CaCl}_2}{110.98 \text{ g CaCl}_2} = 0.1036 \text{ mol CaCl}_2$

   $M = \dfrac{0.1036 \text{ mol}}{0.125 \text{ L}} = 0.829\ M$

38. molar mass of HCHO = 30.026 g; 113.1. mL = 0.1131 L

Chapter 15: Solutions

$$0.1131 \text{ L} \times \frac{3.0 \text{ mol HCHO}}{1 \text{ L}} \times \frac{30.026 \text{ g HCHO}}{1 \text{ mol HCHO}} = 10. \text{ g HCHO}$$

40. molar mass of $I_2$ = 253.8 g; 225 mL = 0.225 L

$$5.15 \text{ g } I_2 \times \frac{1 \text{ mol}}{253.8 \text{ g}} = 0.0203 \text{ mol } I_2$$

$$M = \frac{0.0203 \text{ mol } I_2}{0.225 \text{ L solution}} = 0.0902 \ M$$

42. molar mass of $AgNO_3$ = 169.91 g; 250. mL = 0.2500 L

$$0.2500 \text{ L} \times \frac{0.100 \text{ mol}}{1 \text{ L}} \times \frac{169.91 \text{ g}}{1 \text{ mol}} = 4.25 \text{ g AgNO}_3$$

44. a. 12.5 mL = 0.0125 L

$$0.0125 \text{ L solution} \times \frac{0.104 \text{ mol HCl}}{1.00 \text{ L solution}} = 0.00130 \text{ mol HCl}$$

b. 27.3 mL = 0.0273 L

$$0.0273 \text{ L solution} \times \frac{0.223 \text{ mol NaOH}}{1.00 \text{ L solution}} = 0.00609 \text{ mol NaOH}$$

c. 36.8 mL = 0.0368 L

$$0.0368 \text{ L solution} \times \frac{0.501 \text{ mol HNO}_3}{1.00 \text{ L solution}} = 0.0184 \text{ mol HNO}_3$$

d. 47.5 mL = 0.0475 L

$$0.0475 \text{ L solution} \times \frac{0.749 \text{ mol KOH}}{1.00 \text{ L solution}} = 0.0356 \text{ mol KOH}$$

46. a. molar mass of $CaCl_2$ = 110.98 g; 17.8 mL = 0.0178 L

$$0.0178 \text{ L solution} \times \frac{0.119 \text{ mol CaCl}_2}{1 \text{ L solution}} \times \frac{110.98 \text{ g CaCl}_2}{1 \text{ mol CaCl}_2} = 0.235 \text{ g CaCl}_2$$

b. molar mass of KCl = 74.55 g; 27.6 mL = 0.0276 L

$$0.0276 \text{ L solution} \times \frac{0.288 \text{ mol KCl}}{1 \text{ L solution}} \times \frac{74.55 \text{ g KCl}}{1 \text{ mol KCl}} = 0.593 \text{ g KCl}$$

c. molar mass of $FeCl_3$ = 162.2 g; 35.4 mL = 0.0354 L

$$0.0354 \text{ L solution} \times \frac{0.399 \text{ mol FeCl}_3}{1 \text{ L solution}} \times \frac{162.2 \text{ g FeCl}_3}{1 \text{ mol FeCl}_3} = 2.29 \text{ g FeCl}_3$$

d. molar mass $KNO_3$ = 101.11 g; 46.1 mL = 0.0461 L

$$0.0461 \text{ L solution} \times \frac{0.559 \text{ mol KNO}_3}{1 \text{ L solution}} \times \frac{101.11 \text{ g KNO}_3}{1 \text{ mol KNO}_3} = 2.61 \text{ g KNO}_3$$

Chapter 15: Solutions

48. molar mass of KBr = 119.0 g; 225 mL = 0.225 L

$$0.225 \text{ L solution} \times \frac{0.355 \text{ mol KBr}}{1 \text{ L solution}} \times \frac{119.0 \text{ g KBr}}{1 \text{ mol KBr}} = 9.51 \text{ g KBr}$$

50. a. 10.2 mL = 0.0102 L

$$0.0102 \text{ L} \times \frac{0.451 \text{ mol AlCl}_3}{1.00 \text{ L}} \times \frac{1 \text{ mol Al}^{3+}}{1 \text{ mol AlCl}_3} = 4.60 \times 10^{-3} \text{ mol Al}^{3+}$$

$$0.0102 \text{ L} \times \frac{0.451 \text{ mol AlCl}_3}{1.00 \text{ L}} \times \frac{3 \text{ mol Cl}^-}{1 \text{ mol AlCl}_3} = 1.38 \times 10^{-2} \text{ mol Cl}^-$$

b. 
$$5.51 \text{ L} \times \frac{0.103 \text{ mol Na}_3\text{PO}_4}{1.00 \text{ L}} \times \frac{3 \text{ mol Na}^+}{1 \text{ mol Na}_3\text{PO}_4} = 1.70 \text{ mol Na}^+$$

$$5.51 \text{ L} \times \frac{0.103 \text{ mol Na}_3\text{PO}_4}{1.00 \text{ L}} \times \frac{1 \text{ mol PO}_4^{3-}}{1 \text{ mol Na}_3\text{PO}_4} = 0.568 \text{ mol PO}_4^{3-}$$

c. 1.75 mL = 0.00175 L

$$0.00175 \text{ L} \times \frac{1.25 \text{ mol CuCl}_2}{1.00 \text{ L}} \times \frac{1 \text{ mol Cu}^{2+}}{1 \text{ mol CuCl}_2} = 2.19 \times 10^{-3} \text{ mol Cu}^{2+}$$

$$0.00175 \text{ L} \times \frac{1.25 \text{ mol CuCl}_2}{1.00 \text{ L}} \times \frac{2 \text{ mol Cl}^-}{1 \text{ mol CuCl}_2} = 4.38 \times 10^{-3} \text{ mol Cl}^-$$

d. 25.2 mL = 0.0252 L

$$0.0252 \text{ L} \times \frac{0.00157 \text{ mol Ca(OH)}_2}{1.00 \text{ L}} \times \frac{1 \text{ mol Ca}^{2+}}{1 \text{ mol Ca(OH)}_2} = 3.96 \times 10^{-5} \text{ mol Ca}^{2+}$$

$$0.0252 \text{ L} \times \frac{0.00157 \text{ mol Ca(OH)}_2}{1.00 \text{ L}} \times \frac{2 \text{ mol OH}^-}{1 \text{ mol Ca(OH)}_2} = 7.91 \times 10^{-5} \text{ mol OH}^-$$

52. 240. mL = 0.240 L

$$0.240 \text{ L} \times \frac{0.100 \text{ mol Cl}^-}{1 \text{ L}} \times \frac{1 \text{ mol CaCl}_2}{2 \text{ mol Cl}^-} \times \frac{1 \text{ L}}{0.300 \text{ mol CaCl}_2} = 0.0400 \text{ L CaCl}_2$$

0.0400 L CaCl$_2$ = 40.0 mL CaCl$_2$

54. one third

56. $M_1 \times V_1 = M_2 \times V_2$

a. $M_1 = 0.251$ M  $\qquad$  $M_2 = ?$

$V_1 = 125$ mL  $\qquad$ $V_2 = (125 + 250.) = 375$ mL

$$M_2 = \frac{(0.251 \text{ M})(125 \text{ mL})}{375 \text{ mL}} = 0.0837 \text{ M}$$

Chapter 15: Solutions

b.  $M_1 = 0.499\ M$      $M_2 = ?$
    $V_1 = 445\ \text{mL}$      $V_2 = (445 + 250.) = 695\ \text{mL}$

$$M_2 = \frac{(0.499\ M)(445\ \text{mL})}{695\ \text{mL}} = 0.320\ M$$

c.  $M_1 = 0.101\ M$      $M_2 = ?$
    $V_1 = 5.25\ \text{L}$      $V_2 = (5.25 + 0.250) = 5.50\ \text{L}$

$$M_2 = \frac{(0.101\ M)(5.25\ \text{L})}{5.50\ \text{L}} = 0.0964\ M$$

d.  $M_1 = 14.5\ M$      $M_2 = ?$
    $V_1 = 11.2\ \text{mL}$      $V_2 = (11.2 + 250.) = 261.2\ \text{mL}$

$$M_2 = \frac{(14.5\ M)(11.2\ \text{mL})}{261.2\ \text{mL}} = 0.622\ M$$

58. $M_1 = 19.4\ M$      $M_2 = 3.00\ M$
    $V_1 = ?\ \text{mL}$      $V_2 = 3.50\ \text{L}$

$$M_1 = \frac{(3.00\ M)(3.50\ \text{mL})}{(19.4\ M)} = 0.541\ \text{L}\ (541\ \text{mL})$$

60. $M_1 \times V_1 = M_2 \times V_2$

    $M_1 = 1.01\ M$      $M_2 = 0.150\ M$
    $V_1 = ?\ \text{mL}$      $V_2 = 325\ \text{mL}$

$$M_2 = \frac{(0.150\ M)(325\ \text{mL})}{(1.01\ M)} = 48.3\ \text{mL}$$

Dilute 48.3 mL of the 1.01 $M$ solution to a final volume of 325 mL.

62. $M_1 \times V_1 = M_2 \times V_2$

    $M_1 = 6.00\ M$      $M_2 = 0.300\ M$
    $V_1 = 10.0\ \text{mL} = 0.0100\ \text{L}$      $V_2 = ?$

$$V_2 = \frac{(6.00\ M)(0.0100\ \text{L})}{(0.300\ M)} = 0.200\ \text{L} = 200.\ \text{mL}$$

Therefore 200. − 10.0 = 190. mL of water must be added.

64. $Na_2CO_3(aq) + CaCl_2(aq) \rightarrow CaCO_3(s) + 2NaCl(s)$

mmol Ca$^{2+}$ ion: $37.2\ \text{mL} \times \dfrac{0.105\ \text{mmol Ca}^{2+}}{1.00\ \text{mL}} = 3.91\ \text{mmol Ca}^{2+}$

From the balanced chemical equation, 3.91 mmol $CO_3^{2-}$ will be needed to precipitate this quantity of Ca$^{2+}$ ion.

Chapter 15: Solutions

$$3.91 \text{ mmol CO}_3^{2-} \times \frac{1.00 \text{ mL}}{0.125 \text{ mmol}} = 31.2 \text{ mL}$$

66.  molar mass $Na_2C_2O_4$ = 134.0 g      37.5 mL = 0.0375 L

$$\text{moles Ca}^{2+} \text{ ion} = 0.0375 \text{ L} \times \frac{0.104 \text{ mol Ca}^{2+}}{1.00 \text{ L}} = 0.00390 \text{ mol Ca}^{2+} \text{ ion}$$

$Ca^{2+}(aq) + C_2O_4^{2-}(aq) \rightarrow CaC_2O_4(s)$

As the precipitation reaction is of 1:1 stoichiometry, then 0.00390 mol of $C_2O_4^{2-}$ ion is needed. Moreover, each formula unit of $Na_2C_2O_4$ contains one $C_2O_4^{2-}$ ion, so 0.00390 mol of $Na_2C_2O_4$ is required.

$$0.00390 \text{ mol Na}_2\text{C}_2\text{O}_4 \times \frac{134.0 \text{ g}}{1 \text{ mol}} = 0.523 \text{ g Na}_2\text{C}_2\text{O}_4 \text{ required}$$

68.  10.0 mL = 0.0100 L

$$0.0100 \text{ L} \times \frac{0.250 \text{ mol AlCl}_3}{1.00 \text{ L}} = 2.50 \times 10^{-3} \text{ mol AlCl}_3$$

$AlCl_3(aq) + 3NaOH(s) \rightarrow Al(OH)_3(s) + 3NaCl(aq)$

$$2.50 \times 10^{-3} \text{ mol AlCl}_3 \times \frac{3 \text{ mol NaOH}}{1 \text{ mol AlCl}_3} = 7.50 \times 10^{-3} \text{ mol NaOH}$$

molar mass NaOH = 40.00 g

$$7.50 \times 10^{-3} \text{ mol NaOH} \times \frac{40.00 \text{ g NaOH}}{1 \text{ mol}} = 0.300 \text{ g NaOH}$$

70.  $H_2SO_4(aq) + 2NaOH(aq) \rightarrow 2H_2O(l) + Na_2SO_4(aq)$

40.0 mL = 0.0400 L

$$0.0400 \text{ L} \times \frac{0.400 \text{ mol H}_2\text{SO}_4}{1 \text{ L}} = 0.0160 \text{ mol H}_2\text{SO}_4$$

$$0.0160 \text{ mol H}_2\text{SO}_4 \times \frac{2 \text{ mol NaOH}}{1 \text{ mol H}_2\text{SO}_4} = 0.0320 \text{ mol NaOH}$$

$$0.0320 \text{ mol NaOH} \times \frac{1 \text{ L}}{0.500 \text{ mol NaOH}} = 0.0640 \text{ L NaOH}$$

0.064 L NaOH = 64.0 mL NaOH

72.  7.2 mL = 0.0072 L

$$0.0072 \text{ L} \times \frac{2.5 \times 10^{-3} \text{ mol NaOH}}{1.00 \text{ L}} = 1.8 \times 10^{-5} \text{ mol NaOH}$$

$H^+(aq) + OH^-(aq) \rightarrow H_2O(l)$

Chapter 15: Solutions

$$1.8 \times 10^{-5} \text{ mol OH}^- \times \frac{1 \text{ mol H}^+}{1 \text{ mol OH}^-} = 1.8 \times 10^{-5} \text{ mol H}^+$$

$100 \text{ mL} = 0.100 \text{ L}$

$$M = \frac{1.8 \times 10^{-5} \text{ mol H}^+}{0.100 \text{ L}} = 1.8 \times 10^{-4} \, M \text{ H}^+(aq)$$

74. Experimentally, neutralization reactions are usually performed with volumetric glassware that is calibrated in milliliters rather than liters. For convenience in calculations for such reactions, the arithmetic is often performed in terms of *milli*liters and *milli*moles, rather than in liters and moles: 1 mmol = 0.001 mol. Note that the number of moles of solute per liter of solution, the molarity, is numerically equivalent to the number of *milli*moles of solute per *milli*liter of solution.

   a.  $HNO_3(aq) + NaOH(aq) \rightarrow NaNO_3(aq) + H_2O(l)$

   $$12.7 \text{ mL} \times \frac{0.501 \text{ mmol}}{1.00 \text{ mL}} = 6.36 \text{ mmol NaOH present in the sample}$$

   $$6.36 \text{ mmol NaOH} \times \frac{1 \text{ mmol HNO}_3}{1 \text{ mmol NaOH}} = 6.36 \text{ mmol HNO}_3 \text{ required to react}$$

   $$6.36 \text{ mmol HNO}_3 \times \frac{1.00 \text{ mL}}{0.101 \text{ mmol HNO}_3} = 63.0 \text{ mL HNO}_3 \text{ required}$$

   b.  $2HNO_3(aq) + Ba(OH)_2 \rightarrow Ba(NO_3)_2 + 2H_2O(l)$

   $$24.9 \text{ mL} \times \frac{0.00491 \text{ mmol}}{1.00 \text{ mL}} = 0.122 \text{ mmol Ba(OH)}_2 \text{ present in sample}$$

   $$0.122 \text{ mmol Ba(OH)}_2 \times \frac{2 \text{ mmol HNO}_3}{1 \text{ mmol Ba(OH)}_2} = 0.244 \text{ mmol HNO}_3 \text{ required}$$

   $$0.244 \text{ mmol HNO}_3 \times \frac{1.00 \text{ mL}}{0.101 \text{ mmol HNO}_3} = 2.42 \text{ mL HNO}_3 \text{ is required}$$

   c.  $HNO_3(aq) + NH_3(aq) \rightarrow NH_4NO_3(aq)$

   $$49.1 \text{ mL} \times \frac{0.103 \text{ mmol}}{1.00 \text{ mL}} = 5.06 \text{ mmol NH}_3 \text{ present in the sample}$$

   $$5.06 \text{ mmol NH}_3 \times \frac{1 \text{ mmol HNO}_3}{1 \text{ mmol NH}_3} = 5.06 \text{ mmol HNO}_3 \text{ required}$$

   $$5.06 \text{ mmol HNO}_3 \times \frac{1.00 \text{ mL}}{0.101 \text{ mmol HNO}_3} = 50.1 \text{ mL HNO}_3 \text{ required}$$

   d.  $KOH(aq) + HNO_3(aq) \rightarrow KNO_3(aq) + H_2O(l)$

   $$1.21 \text{ L} \times \frac{0.102 \text{ mol}}{1.00 \text{ L}} = 0.123 \text{ mol KOH present in the sample}$$

   $$0.123 \text{ mol KOH} \times \frac{1 \text{ mol HNO}_3}{1 \text{ mol KOH}} = 0.123 \text{ mol HNO}_3 \text{ required}$$

Chapter 15: Solutions

$$0.123 \text{ mol HNO}_3 \times \frac{1.00 \text{ L}}{0.101 \text{ mol HNO}_3} = 1.22 \text{ L HNO}_3 \text{ required}$$

76. 1 normal

78. 1.53 equivalents OH⁻ ion are needed to react with 1.53 equivalents of H⁺ ion. By *definition*, one equivalent of OH⁻ ion exactly neutralizes one equivalent of H⁺ ion.

80. $N = \dfrac{\text{number of equivalents of solute}}{\text{number of liters of solution}}$

    a. equivalent weight NaOH = molar mass NaOH = 40.00 g

$$0.113 \text{ g NaOH} \times \frac{1 \text{ equiv NaOH}}{40.00 \text{ g}} = 2.83 \times 10^{-3} \text{ equiv NaOH}$$

10.2 mL = 0.0102 L

$$N = \frac{2.83 \times 10^{-3} \text{ equiv}}{0.0102 \text{ L}} = 0.277 \, N$$

    b. equivalent weight Ca(OH)₂ $\dfrac{\text{molar mass}}{2} = \dfrac{74.10 \text{ g}}{2} = 37.05 \text{ g}$

$$12.5 \text{ mg} \times \frac{1 \text{ g}}{10^3 \text{ mg}} \times \frac{1 \text{ equiv}}{37.05 \text{ g}} = 3.37 \times 10^{-4} \text{ equiv Ca(OH)}_2$$

100. mL = 0.100 L

$$N = \frac{3.37 \times 10^{-4} \text{ equiv}}{0.100 \text{ L}} = 3.37 \times 10^{-3} \, N$$

    c. equivalent weight H₂SO₄ = $\dfrac{\text{molar mass}}{2} = \dfrac{98.09 \text{ g}}{2} = 49.05 \text{ g}$

$$12.4 \text{ g} \times \frac{1 \text{ equiv}}{49.05 \text{ g}} = 0.253 \text{ equiv H}_2\text{SO}_4$$

155 mL = 0.155 L

$$N = \frac{0.253 \text{ equiv}}{0.155 \text{ L}} = 1.63 \, N$$

82.   a. $0.134 \, M \text{ NaOH} \times \dfrac{1 \text{ equiv NaOH}}{1 \text{ mol NaOH}} = 0.134 \, N \text{ NaOH}$

    b. $0.00521 \, M \text{ Ca(OH)}_2 \times \dfrac{2 \text{ equiv Ca(OH)}_2}{1 \text{ mol Ca(OH)}_2} = 0.0104 \, N \text{ Ca(OH)}_2$

    c. $4.42 \, M \text{ H}_3\text{PO}_4 \times \dfrac{3 \text{ equiv H}_3\text{PO}_4}{1 \text{ mol H}_3\text{PO}_4} = 13.3 \, N \text{ H}_3\text{PO}_4$

Chapter 15: Solutions

84. Molar mass of Ca(OH)$_2$ = 74.10 g

$$5.21 \text{ mg Ca(OH)}_2 \times \frac{1 \text{ g}}{10^3 \text{ mg}} \times \frac{1 \text{ mol}}{74.10 \text{ g}} = 7.03 \times 10^{-5} \text{ mol Ca(OH)}_2$$

1000. mL = 1.000 L (volumetric flask volume: 4 significant figures).

$$M = \frac{7.03 \times 10^{-5} \text{ mol}}{1.000 \text{ } L} = 7.03 \times 10^{-5} \text{ } M \text{ Ca(OH)}_2$$

$$N = 7.03 \times 10^{-5} \text{ } M \text{ Ca(OH)}_2 \times \frac{2 \text{ equiv Ca(OH)}_2}{1 \text{ mol Ca(OH)}_2} = 1.41 \times 10^{-4} \text{ } N \text{ Ca(OH)}_2$$

86. H$_2$SO$_4$(aq) + 2NaOH(aq) → Na$_2$SO$_4$(aq) + 2H$_2$O(l)

$N_{acid} \times V_{acid} = N_{base} \times V_{base}$

(0.104 $N$)($V_{acid}$) = (0.152 $N$)(15.2 mL)

$V_{acid}$ = 22.2 mL

The 0.104 $M$ sulfuric acid solution is *twice as concentrated* as the 0.104 $N$ sulfuric acid solution (1 mole = 2 equivalents), so half as much will be required to neutralize the same quantity of NaOH = 11.1 mL

88. 2NaOH(aq) + H$_2$SO$_4$(aq) → Na$_2$SO$_4$(aq) + 2 H$_2$O(l)

$$27.34 \text{ mL NaOH} \times \frac{0.1021 \text{ mmol}}{1.00 \text{ mL}} = 2.791 \text{ mmol NaOH}$$

$$2.791 \text{ mmol NaOH} \times \frac{1 \text{ mmol H}_2\text{SO4}}{2 \text{ mmol NaOH}} = 1.396 \text{ mmol H}_2\text{SO}_4$$

$$M = \frac{1.396 \text{ mmol H}_2\text{SO4}}{25.00 \text{ mL}} = 0.05583 \text{ } M \text{ H}_2\text{SO}_4 = 0.1117 \text{ } N \text{ H}_2\text{SO}_4$$

90. Molarity is defined as the number of moles of solute contained in 1 liter of *total* solution volume (solute plus solvent after mixing). In the first case, where 50. g of NaCl is dissolved in 1.0 L of water, the total volume after mixing is *not* known and the molarity cannot be calculated. In the second example, the final volume after mixing is known and the molarity can be calculated simply.

92. $$75 \text{ g solution} \times \frac{25 \text{ g NaCl}}{100. \text{ g solution}} = 18.75 \text{ g NaCl}$$

$$\text{new \%} = \frac{18.75 \text{ g NaCl}}{575 \text{ g solution}} \times 100 = 3.26 = 3.3 \text{ \%}$$

94. molar mass NaHCO$_3$ = 84.01 g; 25.2 mL = 0.0252 L

NaHCO$_3$(s) + HCl(aq) → NaCl(aq) + H$_2$O(l)

$$\text{mol HCl} = \text{mol NaHCO}_3 \text{ required} = 0.0252 \text{ L} \times \frac{6.01 \text{ mol HCl}}{1.00 \text{ L}} = 0.151 \text{ mol}$$

Chapter 15: Solutions

$$0.151 \text{ mol} \times \frac{84.01 \text{ g}}{1 \text{ mol}} = 12.7 \text{ g NaHCO}_3 \text{ required}$$

96. molar mass H$_2$O = 18.0 g

    1.0 L water = 1.0 × 10$^3$ mL water ≅ 1.0 × 10$^3$ g water

    $$1.0 \times 10^3 \text{ g H}_2\text{O} \times \frac{1 \text{ mol H}_2\text{O}}{18.0 \text{ g H}_2\text{O}} = 56 \text{ mol H}_2\text{O}$$

98. 500 mL HCl solution = 0.500 L HCl solution

    $$0.500 \text{ L solution} \times \frac{0.100 \text{ mol HCl}}{1.00 \text{ L solution}} = 0.0500 \text{ mol HCl}$$

    $$0.0500 \text{ mol HCl} \times \frac{22.4 \text{ L}}{1 \text{ mol}} = 1.12 \text{ L HCl gas at STP}$$

100. $$10.0 \text{ g HCl} \times \frac{100. \text{ g solution}}{33.1 \text{ g HCl}} = 30.21 \text{ g solution}$$

    $$30.21 \text{ g solution} \times \frac{1.00 \text{ mL solution}}{1.147 \text{ g solution}} = 26.3 \text{ mL solution}$$

102. 225.0 mL = 0.2250 L    150.0 mL = 0.1500 L

    Determine the total number of moles of HCl in the solution and the total volume of solution.

    $$0.2250 \text{ L solution} \times \frac{2.5 \text{ mol HCl}}{1 \text{ L solution}} = 0.56 \text{ mol HCl}$$

    $$0.1500 \text{ L solution} \times \frac{0.75 \text{ mol HCl}}{1 \text{ L solution}} = 0.11 \text{ mol HCl}$$

    Total number of moles of HCl = 0.56 mol + 0.11 mol = 0.67 mol HCl
    Total volume of solution = 0.2250 L + 0.1500 L = 0.3750 L solution

    $$\text{Molarity} = \frac{\text{moles of solute}}{\text{liters of solution}} = \frac{0.67 \text{ mol HCl}}{0.3750 \text{ L solution}} = 1.8 \, M$$

104. a. $$\frac{x \text{ g NaCl}}{11.5 \text{ g total}} \times 100 = 6.25\% \text{ NaCl}$$

    $x$ = 0.719 g NaCl

    b. $$\frac{x \text{ g NaCl}}{6.25 \text{ g total}} \times 100 = 11.5\% \text{ NaCl}$$

    $x$ = 0.719 g NaCl

    c. $$\frac{x \text{ g NaCl}}{54.3 \text{ g total}} \times 100 = 0.91\% \text{ NaCl}$$

    $x$ = 0.49 g NaCl

Chapter 15: Solutions

d. $\dfrac{x \text{ g NaCl}}{452 \text{ g total}} \times 100 = 12.3\%$ NaCl

$x = 55.6$ g NaCl

106. $\%\text{C} = \dfrac{5.0 \text{ g C}}{(5.0 \text{ g C} + 1.5 \text{ g Ni} + 100. \text{ g Fe})} \times 100 = 4.7\%$ C

$\%\text{Ni} = \dfrac{1.5 \text{ g Ni}}{(5.0 \text{ g C} + 1.5 \text{ g Ni} + 100. \text{ g Fe})} \times 100 = 1.4\%$ Ni

$\%\text{Fe} = \dfrac{100. \text{ g Fe}}{(5.0 \text{ g C} + 1.5 \text{ g Ni} + 100. \text{ g Fe})} \times 100 = 93.9\%$ Fe

108. To say that the solution is 5.5% by mass $Na_2CO_3$ means that 5.5 g of $Na_2CO_3$ are contained in every 100 g of the solution.

500. g solution $\times \dfrac{5.5 \text{ g Na}_2\text{CO}_3}{100. \text{ g solution}} = 28$ g $Na_2CO_3$

110. For NaCl: 125 g solution $\times \dfrac{7.5 \text{ g NaCl}}{100. \text{ g solution}} = 9.4$ g NaCl

For KBr: 125 g solution $\times \dfrac{2.5 \text{ g KBr}}{100. \text{ g solution}} = 3.1$ g KBr

112. 60.0 mL = 0.0600 L

0.0600 L $\times \dfrac{2.00 \text{ mol CaCl}_2}{1 \text{ L}} = 0.120$ mol $CaCl_2$

Adding 40.0 mL of water to the solution will not change the number of moles of $CaCl_2$ (the solute). It will change the concentration, but not the number of moles.

114. (e); molar mass KCl = 74.55 g; molar mass $KClO_3$ = 122.55 g; molar mass $K_3PO_4$ = 212.27 g; molar mass $KNO_3$ = 101.11 g; molar mass $K_2CO_3$ = 138.21 g

The greater the number of *moles* of $K^+$ ions, the greater the actual *number* of $K^+$ ions due to Avogadro's relationship (1 mole = $6.022 \times 10^{23}$ ions). Thus, determining the number of moles of $K^+$ ions for each substance will tell us the greatest number of potassium ions that are dissolved in water.

1.00 g KCl $\times \dfrac{1 \text{ mol KCl}}{74.55 \text{ g KCl}} \times \dfrac{1 \text{ mol K}^+}{1 \text{ mol KCl}} = 0.0134$ mol $K^+$

1.00 g $KClO_3 \times \dfrac{1 \text{ mol KClO}_3}{122.55 \text{ g KClO}_3} \times \dfrac{1 \text{ mol K}^+}{1 \text{ mol KClO}_3} = 0.00816$ mol $K^+$

1.00 g $K_3PO_4 \times \dfrac{1 \text{ mol K}_3\text{PO}_4}{212.27 \text{ g K}_3\text{PO}_4} \times \dfrac{3 \text{ mol K}^+}{1 \text{ mol K}_3\text{PO}_4} = 0.0141$ mol $K^+$

$$1.00 \text{ g KNO}_3 \times \frac{1 \text{ mol KNO}_3}{101.11 \text{ g KNO}_3} \times \frac{1 \text{ mol K}^+}{1 \text{ mol KNO}_3} = 0.00989 \text{ mol K}^+$$

$$1.00 \text{ g K}_2\text{CO}_3 \times \frac{1 \text{ mol K}_2\text{CO}_3}{138.21 \text{ g K}_2\text{CO}_3} \times \frac{2 \text{ mol K}^+}{1 \text{ mol K}_2\text{CO}_3} = 0.0145 \text{ mol K}^+$$

116. a. The balanced equation is:

    $Zn(NO_3)_2(aq) + 2KOH(aq) \rightarrow Zn(OH)_2(s) + 2KNO_3(aq)$

    Determine the number of moles of each reactant present.

    200.0 mL = 0.2000 L       100.0 mL = 0.1000 L

    $$0.2000 \text{ L} \times \frac{0.10 \text{ mol Zn(NO}_3)_2}{1 \text{ L}} = 0.020 \text{ mol Zn(NO}_3)_2$$

    $$0.1000 \text{ L} \times \frac{0.10 \text{ mol KOH}}{1 \text{ L}} = 0.010 \text{ mol KOH}$$

    Determine the limiting reactant.

    $$0.010 \text{ mol KOH} \times \frac{1 \text{ mol Zn(NO}_3)_2}{2 \text{ mol KOH}} = 0.0050 \text{ mol Zn(NO}_3)_2 \text{ needed}$$

    Since 0.0050 mol Zn(NO₃)₂ is needed to react with all 0.010 mol KOH, and there is 0.020 mol Zn(NO₃)₂ available, KOH will run out first. Thus, KOH is the limiting reactant and will determine how much precipitate, Zn(OH)₂, is produced.

    $$0.010 \text{ mol KOH} \times \frac{1 \text{ mol Zn(OH)}_2}{2 \text{ mol KOH}} = 0.0050 \text{ mol Zn(OH)}_2 \text{ produced}$$

    b. 0.020 mol Zn(NO₃)₂ to start − 0.0050 mol Zn(NO₃)₂ used up = 0.015 mol Zn(NO₃)₂ left

    $$0.015 \text{ mol Zn(NO}_3)_2 \times \frac{1 \text{ mol Zn}^{2+}}{1 \text{ mol Zn(NO}_3)_2} = 0.015 \text{ mol Zn}^{2+} \text{ leftover}$$

    Total volume after the reaction = 0.2000 L + 0.1000 L = 0.3000 L

    $$\frac{0.015 \text{ mol Zn}^{2+}}{0.3000 \text{ L}} = 0.050 \; M \; \text{Zn}^{2+}$$

118. a. $$4.25 \text{ L solution} \times \frac{0.105 \text{ mol KCl}}{1.00 \text{ L solution}} = 0.446 \text{ mol KCl}$$

    molar mass KCl = 74.6 g

    $$0.446 \text{ mol KCl} \times \frac{74.6 \text{ g KCl}}{1 \text{ mol KCl}} = 33.3 \text{ g KCl}$$

    b. 15.1 mL = 0.0151 L

    $$0.0151 \text{ L solution} \times \frac{0.225 \text{ mol NaNO}_3}{1.00 \text{ L solution}} = 3.40 \times 10^{-3} \text{ mol NaNO}_3$$

Chapter 15: Solutions

molar mass NaNO$_3$ = 85.00 g

$3.40 \times 10^{-3}$ mol $\times \dfrac{85.00 \text{ g NaNO}_3}{1 \text{ mol NaNO}_3}$ = 0.289 g NaNO$_3$

c. 25 mL = 0.025 L

0.025 L solution $\times \dfrac{3.0 \text{ mol HCl}}{1.00 \text{ L solution}}$ = 0.075 mol HCl

molar mass HCl = 36.46 g

0.075 mol HCl $\times \dfrac{36.46 \text{ g HCl}}{1 \text{ mol HCl}}$ = 2.7 g HCl

d. 100. mL = 0.100 L

0.100 L solution $\times \dfrac{0.505 \text{ mol H}_2\text{SO}_4}{1.00 \text{ L solution}}$ = 0.0505 mol H$_2$SO$_4$

molar mass H$_2$SO$_4$ = 98.09 g

0.0505 mol H$_2$SO$_4$ $\times \dfrac{98.09 \text{ g H}_2\text{SO}_4}{1 \text{ mol H}_2\text{SO}_4}$ = 4.95 g H$_2$SO$_4$

120. a. 1.25 L $\times \dfrac{0.250 \text{ mol Na}_3\text{PO}_4}{1.00 \text{ L}}$ = 0.3125 mol Na$_3$PO$_4$

0.3125 mol Na$_3$PO$_4$ $\times \dfrac{3 \text{ mol Na}^+}{1 \text{ mol Na}_3\text{PO}_4}$ = 0.938 mol Na$^+$

0.3125 mol Na$_3$PO$_4$ $\times \dfrac{1 \text{ mol PO}_4^-}{1 \text{ mol Na}_3\text{PO}_4}$ = 0.313 mol PO$_4^{3-}$

b. 3.5 mL = 0.0035 L

0.0035 L $\times \dfrac{6.0 \text{ mol H}_2\text{SO}_4}{1.00 \text{ L}}$ = 0.021 mol H$_2$SO$_4$

0.021 mol H$_2$SO$_4$ $\times \dfrac{2 \text{ mol H}^+}{1 \text{ mol H}_2\text{SO}_4}$ = 0.042 mol H$^+$

0.021 mol H$_2$SO$_4$ $\times \dfrac{1 \text{ mol SO}_4^{2-}}{1 \text{ mol H}_2\text{SO}_4}$ = 0.021 mol SO$_4^{2-}$

c. 25 mL = 0.025 L

0.025 L $\times \dfrac{0.15 \text{ mol AlCl}_3}{1.00 \text{ L}}$ = 0.00375 mol AlCl$_3$

0.00375 mol AlCl$_3$ $\times \dfrac{1 \text{ mol Al}^{3+}}{1 \text{ mol AlCl}_3}$ = 0.0038 mol Al$^{3+}$

Chapter 15: Solutions

$$0.00375 \text{ mol AlCl}_3 \times \frac{1 \text{ mol Cl}^-}{1 \text{ mol AlCl}_3} = 0.011 \text{ mol Cl}^-$$

d. $1.50 \text{ L} \times \dfrac{1.25 \text{ mol BaCl}_2}{1.00 \text{ L}} = 1.875 \text{ mol BaCl}_2$

$$1.875 \text{ mol BaCl}_2 \times \frac{1 \text{ mol Ba}^{2+}}{1 \text{ mol BaCl}_2} = 1.88 \text{ mol Ba}^{2+}$$

$$1.875 \text{ mol BaCl}_2 \times \frac{2 \text{ mol Cl}^-}{1 \text{ mol BaCl}_2} = 3.75 \text{ mol Cl}^-$$

122. $M_1 \times V_1 = M_2 \times V_2$

    a.    $M_1 = 0.200 \text{ M}$      $M_2 = ?$

        $V_1 = 125 \text{ mL}$      $V_2 = 125 + 150. = 275 \text{ mL}$

$$M_2 = \frac{(0.200 \text{ M})(125 \text{ mL})}{(275 \text{ mL})} = 0.0909 \text{ M}$$

    b.    $M_1 = 0.250 \text{ M}$      $M_2 = ?$

        $V_1 = 155 \text{ mL}$      $V_2 = 155 + 150. = 305 \text{ mL}$

$$M_2 = \frac{(0.250 \text{ M})(155 \text{ mL})}{(305 \text{ mL})} = 0.127 \text{ M}$$

    c.    $M_1 = 0.250 \text{ M}$      $M_2 = ?$

        $V_1 = 0.500 \text{ L} = 500. \text{ mL}$      $V_2 = 500. + 150. = 650. \text{ mL}$

$$M_2 = \frac{(0.250 \text{ M})(500. \text{ mL})}{(650 \text{ mL})} = 0.192 \text{ M}$$

    d.    $M_1 = 18.0 \text{ M}$      $M_2 = ?$

        $V_1 = 15 \text{ mL}$      $V_2 = 15 + 150. = 165 \text{ mL}$

$$M_2 = \frac{(18.0 \text{ M})(15 \text{ mL})}{(165 \text{ mL})} = 1.6 \text{ M}$$

124. (b);

The balanced equation is: $2\text{AgNO}_3(aq) + \text{CuCl}_2(aq) \rightarrow 2\text{AgCl}(s) + \text{Cu(NO}_3)_2(aq)$

Determine the number of moles of each reactant present.

$$1.0 \text{ L} \times \frac{3.0 \text{ mol AgNO}_3}{1 \text{ L}} = 3.0 \text{ mol AgNO}_3$$

$$1.0 \text{ L} \times \frac{1.0 \text{ mol CuCl}_2}{1 \text{ L}} = 1.0 \text{ mol CuCl}_2$$

Determine the limiting reactant.

Chapter 15: Solutions

$$3.0 \text{ mol AgNO}_3 \times \frac{2 \text{ mol AgCl}}{2 \text{ mol AgNO}_3} = 3.0 \text{ mol AgCl}$$

$$1.0 \text{ mol CuCl}_2 \times \frac{2 \text{ mol AgCl}}{1 \text{ mol CuCl}_2} = 2.0 \text{ mol AgCl}$$

The maximum amount of AgCl that can be produced is 2.0 mol, and then CuCl$_2$ is completely consumed (and is therefore the limiting reactant).

Adding more AgNO$_3$ (as specified in *a*) will not increase the amount of AgCl produced since it is not the limiting reactant. Adding more CuCl$_2$ (as specified in *b*) will increase the amount of AgCl produced since it is the limiting reactant. Adding more water to the AgNO$_3$ solution (as specified in *c*) will have no effect since it does not affect the number of moles of AgNO$_3$ present. Allowing the water to evaporate (as specified in *d*) will also have no effect on the amount of AgCl produced since it does not affect the number of moles of solute present.

126. a. *Molecular*: $3(NH_4)_2S(aq) + 2FeCl_3(aq) \rightarrow Fe_2S_3(s) + 6NH_4Cl(aq)$

   *Complete*: $6NH_4^+(aq) + 3S^{2-}(aq) + 2Fe^{3+}(aq) + 6Cl^-(aq) \rightarrow Fe_2S_3(s) + 6NH_4^+(aq) + 6Cl^-(aq)$

   *Net*: $3S^{2-}(aq) + 2Fe^{3+}(aq) \rightarrow Fe_2S_3(s)$

   b. Determine the number of moles of each reactant present.

   50.0 mL = 0.0500 L          100.0 mL = 0.1000 L

   $$0.0500 \text{ L} \times \frac{0.500 \text{ mol (NH}_4)_2\text{S}}{1 \text{ L}} = 0.0250 \text{ mol (NH}_4)_2\text{S}$$

   $$0.1000 \text{ L} \times \frac{0.250 \text{ mol FeCl}_3}{1 \text{ L}} = 0.0250 \text{ mol FeCl}_3$$

   Determine the limiting reactant.

   $$0.0250 \text{ mol (NH}_4)_2\text{S} \times \frac{1 \text{ mol Fe}_2\text{S}_3}{3 \text{ mol (NH}_4)_2\text{S}} = 0.00833 \text{ mol Fe}_2\text{S}_3$$

   $$0.0250 \text{ mol FeCl}_3 \times \frac{1 \text{ mol Fe}_2\text{S}_3}{2 \text{ mol FeCl}_3} = 0.0125 \text{ mol Fe}_2\text{S}_3$$

   The maximum amount of Fe$_2$S$_3$ that can be produced is 0.00833 mol, and then (NH$_4$)$_2$S is completely consumed (and is therefore the limiting reactant).

   molar mass of Fe$_2$S$_3$ = 207.91 g

   $$0.008\overline{3}3 \text{ mol Fe}_2\text{S}_3 \times \frac{207.91 \text{ g Fe}_2\text{S}_3}{1 \text{ mol Fe}_2\text{S}_3} = 1.73 \text{ g Fe}_2\text{S}_3$$

   c. [NH$_4^+$] is a spectator ion and does not participate in the chemical reaction to make the precipitate.

   $$0.0250 \text{ mol (NH}_4)_2\text{S} \times \frac{2 \text{ mol NH}_4^+}{1 \text{ mol (NH}_4)_2\text{S}} = 0.0500 \text{ mol NH}_4^+$$

   Total volume = 0.0500 L + 0.1000 L = 0.1500 L

Chapter 15: Solutions

$$\frac{0.0500 \text{ mol NH}_4^+}{0.1500 \text{ L}} = 0.333 \ M \text{ NH}_4^+$$

[$Fe^{3+}$] is not a spectator ion and does participate in the chemical reaction to make the precipitate and will be in excess since it is not the limiting reactant.

$$0.0250 \text{ mol FeCl}_3 \times \frac{1 \text{ mol Fe}^{3+}}{1 \text{ mol FeCl}_3} = 0.0250 \text{ mol Fe}^{3+} \text{ to start}$$

$$0.0250 \text{ mol (NH}_4)_2\text{S} \times \frac{1 \text{ mol S}^{2-}}{1 \text{ mol (NH}_4)_2\text{S}} = 0.0250 \text{ mol S}^{2-} \text{ to start}$$

Using the net ionic equation: $3S^{2-}(aq) + 2Fe^{3+}(aq) \rightarrow Fe_2S_3(s)$

$$0.0250 \text{ mol S}^{2-} \times \frac{2 \text{ mol Fe}^{3+}}{3 \text{ mol S}^{2-}} = 0.0167 \text{ mol Fe}^{3+} \text{ used up}$$

0.0250 mol $Fe^{3+}$ to start – 0.0167 mol $Fe^{3+}$ used up = 0.0083 mol $Fe^{3+}$ leftover

$$\frac{0.0083 \text{ mol Fe}^{3+}}{0.1500 \text{ L}} = 0.055 \ M \text{ Fe}^{3+}$$

d.  molar mass $FeCl_3$ = 162.2 g

In part b, it was determined that 0.0250 mol $FeCl_3$ was needed to begin the experiment. Convert this to grams (the typical measurement for a solid).

$$0.0250 \text{ mol FeCl}_3 \times \frac{162.2 \text{ g FeCl}_3}{1 \text{ mol FeCl}_3} = 4.06 \text{ g FeCl}_3$$

Measure 4.06 g $FeCl_3$ on a balance, place it into a flask, and dissolve with water until the total volume of the solution is 100.0 mL.

128. $M_1 \times V_1 = M_2 \times V_2$

$M_1 = 16 \ M$ $\qquad\qquad M_2 = 0.10 \ M$

$V_1 = ?$ $\qquad\qquad V_2 = 750 \text{ mL} = 0.75 \text{ L}$

$$V_1 = \frac{(0.10 \ M)(0.75 \text{ L})}{16 \ M} = 0.0047 \text{ L}$$

0.0047 L = 4.7 mL

130. a.  equivalent weight HCl = molar mass HCl = 36.46 g; 500. mL = 0.500 L

$$15.0 \text{ g HCl} \times \frac{1 \text{ equiv HCl}}{36.46 \text{ g HCl}} = 0.411 \text{ equiv HCl}$$

$$N = \frac{0.411 \text{ equiv}}{0.500 \text{ L}} = 0.822 \ N$$

b.  equivalent weight $H_2SO_4$ = $\frac{\text{molar mass}}{2} = \frac{98.09 \text{ g}}{2}$ = 49.05 g; 250. mL = 0.250 L

Chapter 15: Solutions

$$49.0 \text{ g H}_2\text{SO}_4 \times \frac{1 \text{ equiv H}_2\text{SO}_4}{49.05 \text{ g H}_2\text{SO}_4} = 0.999 \text{ equiv H}_2\text{SO}_4$$

$$N = \frac{0.999 \text{ equiv}}{0.250 \text{ L}} = 4.00 \, N$$

c. equivalent weight $H_3PO_4 = \dfrac{\text{molar mass}}{3} = \dfrac{98.0 \text{ g}}{3} = 32.67$ g; 100. mL = 0.100 L

$$10.0 \text{ g H}_3\text{PO}_4 \times \frac{1 \text{ equiv H}_3\text{PO}_4}{32.67 \text{ g H}_3\text{PO}_4} = 0.3061 \text{ equiv H}_3\text{PO}_4$$

$$N = \frac{0.3061 \text{ equiv}}{0.100 \text{ L}} = 3.06 \, N$$

132. molar mass $NaH_2PO_4$ = 120.0 g; 500. mL = 0.500 L

$$5.0 \text{ g NaH}_2\text{PO}_4 \times \frac{1 \text{ mol NaH}_2\text{PO}_4}{120.0 \text{ g NaH}_2\text{PO}_4} = 0.04167 \text{ mol NaH}_2\text{PO}_4$$

$$M = \frac{0.04167 \text{ mol}}{0.500 \text{ L}} = 0.08333 \, M \text{ NaH}_2\text{PO}_4 = 0.083 \, M \text{ NaH}_2\text{PO}_4$$

$$0.08333 \, M \text{ NaH}_2\text{PO}_4 \times \frac{2 \text{ equiv NaH}_2\text{PO}_4}{1 \text{ mol NaH}_2\text{PO}_4} = 0.1667 \, N \text{ NaH}_2\text{PO}_4 = 0.17 \, N \text{ NaH}_2\text{PO}_4$$

134. $N_{acid} \times V_{acid} = N_{base} \times V_{base}$

$N_{acid} \times (10.0 \text{ mL}) = (3.5 \times 10^{-2} \, N)(27.5 \text{ mL})$

$N_{acid} = 9.6 \times 10^{-2} \, N \text{ HNO}_3$

136. The balanced equation is: $Mg(NO_3)_2(aq) + 2KOH(aq) \rightarrow Mg(OH)_2(s) + 2KNO_3(aq)$

   molar mass of $Mg(OH)_2$ = 58.326 g

   a. Determine the number of moles of each reactant present.

$$0.156 \text{ L} \times \frac{0.105 \text{ mol Mg(NO}_3)_2}{1 \text{ L}} = 0.0164 \text{ mol Mg(NO}_3)_2$$

$$0.166 \text{ L} \times \frac{0.106 \text{ mol KOH}}{1 \text{ L}} = 0.0176 \text{ mol KOH}$$

   Determine the limiting reactant.

$$0.0164 \text{ mol Mg(NO}_3)_2 \times \frac{1 \text{ mol Mg(OH)}_2}{1 \text{ mol Mg(NO}_3)_2} = 0.0164 \text{ mol Mg(OH)}_2$$

$$0.0176 \text{ mol KOH} \times \frac{1 \text{ mol Mg(OH)}_2}{2 \text{ mol KOH}} = 0.00880 \text{ mol Mg(OH)}_2$$

   The maximum amount of $Mg(OH)_2$ that can be produced is 0.00880 mol, and then KOH is completely consumed (and is therefore the limiting reactant).

Chapter 15: Solutions

$$0.00880 \text{ mol Mg(OH)}_2 \times \frac{58.326 \text{ g Mg(OH)}_2}{1 \text{ mol Mg(OH)}_2} = 0.513 \text{ g Mg(OH)}_2$$

b.  [NO$_3^-$] is a spectator ion and does not participate in the chemical reaction to make the precipitate.

$$0.0164 \text{ mol Mg(NO}_3)_2 \times \frac{2 \text{ mol NO}_3^-}{1 \text{ mol Mg(NO}_3)_2} = 0.0328 \text{ mol NO}_3^-$$

Total volume = 0.156 L + 0.166 L = 0.322 L

$$\frac{0.0328 \text{ mol NO}_3^-}{0.322 \text{ L}} = 0.102 \ M \text{ NO}_3^-$$

138. molar mass H$_2$C$_2$O$_4$ = 90.036 g

100.0 mL = 0.1000 L; 10.00 mL = 0.01000 L; 250.0 mL = 0.2500 L

$$0.6706 \text{ g H}_2\text{C}_2\text{O}_4 \times \frac{1 \text{ mol H}_2\text{C}_2\text{O}_4}{90.036 \text{ g H}_2\text{C}_2\text{O}_4} = 0.007448 \text{ mol H}_2\text{C}_2\text{O}_4$$

$$M = \frac{0.007448 \text{ mol H}_2\text{C}_2\text{O}_4}{0.1000 \text{ L}} = 0.07448 \ M \text{ H}_2\text{C}_2\text{O}_4$$

$$0.01000 \text{ L} \times \frac{0.07448 \text{ mol H}_2\text{C}_2\text{O}_4}{1 \text{ L}} = 7.448 \times 10^{-4} \text{ mol H}_2\text{C}_2\text{O}_4$$

$$M = \frac{7.448 \times 10^{-4} \text{ mol H}_2\text{C}_2\text{O}_4}{0.2500 \text{ L}} = 2.979 \times 10^{-3} \ M \text{ H}_2\text{C}_2\text{O}_4$$

140. 500.0 mL = 0.5000 L; 400.0 mL = 0.4000 L

molar mass Ba$_3$(PO$_4$)$_2$ = 601.84 g

The balanced equation is: 2Na$_3$PO$_4$(aq) + 3BaCl$_2$(aq) → Ba$_3$(PO$_4$)$_2$(s) + 6NaCl(aq)

$$0.5000 \text{ L} \times \frac{0.200 \text{ mol Na}_3\text{PO}_4}{1 \text{ L}} = 0.100 \text{ mol Na}_3\text{PO}_4$$

$$0.4000 \text{ L} \times \frac{0.289 \text{ mol BaCl}_2}{1 \text{ L}} = 0.116 \text{ mol BaCl}_2$$

BaCl$_2$ is the limiting reactant (only 0.0771 mol Na$_3$PO$_4$ needed).

$$0.116 \text{ mol BaCl}_2 \times \frac{1 \text{ mol Ba}_3(\text{PO}_4)_2}{3 \text{ mol BaCl}_2} = 0.0385 \text{ mol Ba}_3(\text{PO}_4)_2$$

$$0.0385 \text{ mol Ba}_3(\text{PO}_4)_2 \times \frac{601.84 \text{ g Ba}_3(\text{PO}_4)_2}{1 \text{ mol Ba}_3(\text{PO}_4)_2} = 23.2 \text{ g Ba}_3(\text{PO}_4)_2 \text{(when decimals carried over)}$$

Chapter 15: Solutions

142. 34.66 mL = 0.03466 L; 50.00 mL = 0.05000 L

The balanced equation is: $Ca(OH)_2(aq) + 2HNO_3(aq) \rightarrow Ca(NO_3)_2(aq) + 2H_2O(l)$

$0.03466 \text{ L} \times \dfrac{0.944 \text{ mol HNO}_3}{1 \text{ L}} = 0.0327 \text{ mol HNO}_3$

$0.0327 \text{ mol HNO}_3 \times \dfrac{1 \text{ mol Ca(OH)}_2}{2 \text{ mol HNO}_3} = 0.0164 \text{ mol Ca(OH)}_2$

$M = \dfrac{0.0164 \text{ mol Ca(OH)}_2}{0.05000 \text{ L}} = 0.327 \ M \ \text{Ca(OH)}_2$ (when decimals carried over)

# CUMULATIVE REVIEW

# Chapters 13–15

2. The pressure of the atmosphere represents the mass of the gases in the atmosphere pressing down on the surface of the earth. The device most commonly used to measure the pressure of the atmosphere is the mercury barometer shown in Figure 13.2 in the text.

   A simple experiment to demonstrate the pressure of the atmosphere is shown in Figure 13.1 in the text. Some water is added to a metal can, and the can heated until the water boils (boiling represents when the pressure of the vapor coming from the water is equal to the atmospheric pressure). The can is then stoppered. As the steam in the can cools, it condenses to liquid water, which lowers the pressure of gas inside the can. The pressure of the atmosphere outside the can is then much larger than the pressure inside the can, and the can collapses.

4. In simple terms, Boyle's law states that the volume of a gas sample will decrease if you squeeze harder on it. Imagine squeezing hard on a tennis ball with your hand: the ball collapses as the gas inside is forced into a smaller volume by your hand. Of course, to be perfectly correct, the temperature and amount of gas (moles) must remain the same as you adjust the pressure for Boyle's law to hold true. There are two mathematical statements of Boyle's law you should remember. The first is

    $P \times V =$ constant,

   which basically is the definition of Boyle's law (in order for the product ($P \times V$) to remain constant, if one of these terms increases the other must decrease). The second formula is the one more commonly used in solving problems,

    $P_1 \times V_1 = P_2 \times V_2$.

   With this second formulation, we can determine pressure-volume information about a given sample under two sets of conditions. These two mathematical formulas are just two different ways of saying the same thing: if the pressure on a sample of gas is increased, the volume of the sample of gas will decrease. A graph of Boyle's law data is given as Figure 13.5: this type of graph ($xy = k$) is known to mathematicians as a hyperbola.

6. Charles's law simply says that if you heat a sample of gas, the volume of the sample will increase. That is, when the temperature of a gas is increased, the volume of the gas also increases (assuming the pressure and amount of gas remains the same). Charles's law is a direct proportionality when the temperature is expressed in kelvins (if you increase $T$, this increases $V$), whereas Boyle's law is an inverse proportionality (if you increase $P$, this decreases $V$). There are two mathematical statements of Charles's law with which you should be familiar. The first statement is:

    $V = kT$.

   This is simply a definition (the volume of a gas sample is directly related to its Kelvin temperature: if you increase the temperature, the volume increases). The working formulation of Charles's law we use in problem solving is given as:

Review:     Chapters 13, 14, and 15

$$\frac{V_1}{T_1} = \frac{V_2}{T_2}$$

With this formulation, we can determine volume-temperature information for a given gas sample under two sets of conditions. Charles's law only holds true if the amount of gas remains the same (obviously the volume of a gas sample would increase if there were more gas present) and also if the pressure remains the same (a change in pressure also changes the volume of a gas sample).

8. Avogadro's law tells us that, with all other things being equal, two moles of gas are twice as big as one mole of gas! That is, the volume of a sample of gas is directly proportional to the number of moles or molecules of gas present (at constant temperature and pressure). If we want to compare the volumes of two samples of the same gas as an indication of the amount of gas present in the samples, we would have to make certain that the two samples of gas are at the same pressure and temperature: the volume of a sample of gas would vary with either temperature or pressure, or both. Avogadro's law holds true for comparing gas samples that are under the same conditions. Avogadro's law is a direct proportionality: the greater the number of gas molecules you have in a sample, the larger the sample's volume will be.

10. The "partial" pressure of an individual gas in a mixture of gases represents the pressure the gas would have in the same container at the same temperature if it were the only gas present. The total pressure in a mixture of gases is the sum of the individual partial pressures of the gases present in the mixture. Because the partial pressures of the gases in a mixture are additive (i.e., the total pressure is the sum of the partial pressures), this suggests that the total pressure in a container is a function only of the number of molecules present in the same, and not of the identity of the molecules or any other property of the molecules (such as their inherent atomic size).

12. The main postulates of the kinetic-molecular theory for gases are as follows: (a) gases consist of tiny particles (atoms or molecules), and the size of these particles is negligible compared to the bulk volume of a gas sample; (b) the particles in a gas are in constant random motion, colliding with each other and with the walls of the container; (c) the particles in a gas sample do not exert any attractive or repulsive forces on one another; (d) the average kinetic energy of the particles in a sample of gas is directly related to the absolute temperature of the gas sample. The pressure exerted by a gas is a result of the molecules colliding with (and pushing on) the walls of the container. The pressure increases with temperature because at a higher temperature, the molecules are moving faster and hit the walls of the container with greater force. A gas fills whatever volume is available to it because the molecules in a gas are in constant random motion: if the motion of the molecules is random, they eventually will move out into whatever volume is available until the distribution of molecules is uniform. At constant pressure, the volume of a gas sample increases as the temperature is increased because with each collision having greater force, the container must expand so that the molecules (and therefore the collisions) are farther apart if the pressure is to remain constant.

14. Solids and liquids are much more condensed states of matter than are gases: the molecules are much closer together in solids and liquids and interact with each other to a much greater extent. Solids and liquids have much greater densities than do gases, and are much less compressible, because there is so little room between the molecules in the solid and liquid states (solids and liquids effectively have native volumes of their own, and their volumes are not affected nearly as much by the temperature or pressure). Although solids are more rigid than liquids, the solid and liquid state have much more in common with each other than either of these states has with the gaseous state. We know this is true because it typically only takes a few kilojoules of energy to

Review: Chapters 13, 14, and 15

melt 1 mol of a solid (not much change has to take place in the molecules), whereas it may take 10 times more energy to vaporize a liquid (as there is a great change between the liquid and gaseous states).

16. The normal boiling point of water, that is, water's boiling point at a pressure of exactly 760 mm Hg, is 100°C (you will recall that the boiling point of water was used to set one of the reference temperatures of the Celsius temperature scale). Water remains at 100°C while boiling, until all the water has boiled away, because the additional heat energy being added to the sample is used to overcome attractive forces among the water molecules as they go from the condensed, liquid state to the gaseous state. The normal (760 mm Hg) freezing point of water is exactly 0°C (again, this property of water was used as one of the reference points for the Celsius temperature scale). A cooling curve for water is given in Figure 14.2. Notice how the curve shows that the amount of heat needed to boil the sample is much larger than the amount needed to melt the sample.

18. Dipole-dipole forces are a type of intermolecular force that can exist between molecules with permanent dipole moments. Molecules with permanent dipole moments try to orient themselves so that the positive end of one polar molecule can attract the negative end of another polar molecule. Dipole-dipole forces are not nearly as strong as ionic or covalent bonding forces (only about 1% as strong as covalent bonding forces) because electrostatic attraction is related to the magnitude of the charges of the attracting species. As polar molecules have only a "partial" charge at each end of the dipole, the magnitude of the attractive force is not as large. The strength of such forces also drops rapidly as molecules become farther apart and is important only in the solid and liquid states (such forces are negligible in the gaseous state because the molecules are too far apart). Hydrogen bonding is an especially strong sort of dipole-dipole attractive force that can exist when hydrogen atoms are directly bonded to the most strongly electronegative atoms (N, O, and F). Because the hydrogen atom is so small, dipoles involving N–H, O–H, and F–H bonds can approach each other much more closely than can dipoles involving other atoms. As the magnitude of dipole-dipole forces is dependent on distance, unusually strong attractive forces can exist in such molecules. We take the fact that the boiling point of water is higher than that of the other covalent hydrogen compounds of the Group 6 elements as evidence for the special strength of hydrogen bonding (it takes more energy to vaporize water because of the extra strong forces holding together the molecules in the liquid state).

20. Vaporization of a liquid requires an input of energy because the intermolecular forces that hold the molecules together in the liquid state must be overcome. The high heat of vaporization of water is essential to life on Earth because much of the excess energy striking the Earth from the sun is dissipated in vaporizing water. Condensation is the opposite process to vaporization; that is, condensation refers to the process by which molecules in the vapor state form a liquid. In a closed container containing a liquid and some empty space above the liquid, an equilibrium is set up between vaporization and condensation. The liquid in such a sealed container never completely evaporates: when the liquid is first placed in the container, the liquid phase begins to evaporate into the empty space. As the number of molecules in the vapor phase begins to get large, however, some of these molecules begin to re-enter the liquid phase. Eventually, every time a molecule of liquid somewhere in the container enters the vapor phase, somewhere else in the container a molecule of vapor re-enters the liquid. There is no further net change in the amount of liquid phase (although molecules are continually moving between the liquid and vapor phases). The pressure of the vapor in such an equilibrium situation is characteristic for the liquid at each particular temperature (for example, the vapor pressures of water are tabulated at different temperatures in Table 13.2). A simple experiment to determine vapor pressure is shown in Figure 14.10. Samples of a liquid are injected into a sealed tube containing mercury; because mercury is so dense, the liquids float to the top of the mercury where they evaporate. As the vapor pressures

Review: Chapters 13, 14, and 15

of the liquids develop to the saturation point, the level of mercury in the tube changes as an index of the magnitude of the vapor pressures. Typically, liquids with strong intermolecular forces have small vapor pressures (they have more difficulty in evaporating) than do liquids with very weak intermolecular forces: for example, the components of gasoline (weak forces) have much higher vapor pressures, and evaporate more easily than does water (strong forces).

22. The simple model we use to explain many properties of metallic elements is called the electron sea model. In this model we picture a regular lattice array of metal cations in sort of a "sea" of mobile valence electrons. The electrons can move easily to conduct heat or electricity through the metal; and the lattice of cations can be deformed fairly easily, allowing the metal to be hammered into a sheet or stretched to make a wire. An alloy contains a mixture of elements, which overall has metallic properties. Substitutional alloys consist of a host metal in which some of the atoms in the metal's crystalline structure are replaced by atoms of other metallic elements of comparable size to the atoms of the host metal. For example, sterling silver consists of an alloy in which approximately 7% of the silver atoms have been replaced by copper atoms. Brass and pewter are also substitutional alloys. An interstitial alloy is formed when other smaller atoms enter the interstices (holes) between atoms in the host metal's crystal structure. Steel is an interstitial alloy in which typically carbon atoms enter the interstices of a crystal of iron atoms. The presence of the interstitial carbon atoms markedly changes the properties of the iron, making it much harder, more malleable, and more ductile. Depending on the amount of carbon introduced into the iron crystals, the properties of the resulting steel can be carefully controlled.

24. A saturated solution is one that contains as much solute as can dissolve at a particular temperature. To say that a solution is saturated does not necessarily mean that the solute is present at a high concentration. For example, magnesium hydroxide only dissolves to a very small extent before the solution is saturated, whereas it takes a great deal of sugar to form a saturated solution (and the saturated solution is extremely concentrated). A saturated solution is one which is in equilibrium with undissolved solute: as molecules of solute dissolve from the solid in one place in the solution, dissolved molecules rejoin the solid phase in another place in the solution. As with the development of vapor pressure above a liquid (see Question 20 above), formation of a solution reaches a state of dynamic equilibrium: once the rates of dissolving and "undissolving" become equal, there will be no further net change in the concentration of the solution and the solution will be saturated.

26. Adding more solvent to a solution so as to dilute the solution *does not change* the number of moles of solute present, but only changes the volume in which the solute is dispersed. If we are using the molarity of the solution to describe its concentration, the number of liters is changed when we add solvent, and the number of moles per liter (the molarity) changes, but the actual number of moles of solute does not change. For example, 125 mL of 0.551 M NaCl contains 68.9 millimol of NaCl. The solution will still contain 68.9 millimol of NaCl after the 250 mL of water is added to it, only now the 68.9 millimol of NaCl will be dispersed in a total volume of 375 mL. This gives the new molarity as 68.9 mmol/375 mL = 0.184 M. The volume and the concentration have changed, but the number of moles of solute in the solution has not changed.

28. $P_1 \times V_1 = P_2 \times V_2$

a. $V_2 = \dfrac{P_1 \times V_1}{P_2} = \dfrac{125 \text{ mL} \times 755 \text{ mm Hg}}{899 \text{ mm Hg}} = 105 \text{ mL}$

b. $P_2 = \dfrac{P_1 \times V_1}{V_2} = \dfrac{455 \text{ mL} \times 755 \text{ mm Hg}}{327 \text{ mL}} = 1.05 \times 10^3 \text{ mm Hg}$

Review: Chapters 13, 14, and 15

30. a. $PV = nRT$; molar mass He = 4.003 g; 25°C = 298 K

$$1.15 \text{ g He} \times \frac{1 \text{ mol}}{4.003 \text{ g}} = 0.2873 \text{ mol He}$$

$$V = \frac{nRT}{P} = \frac{(0.2873 \text{ mol})(0.08206 \text{ L-atm/mol-K})(298 \text{ K})}{(1.01 \text{ atm})} = 6.96 \text{ L}$$

b. molar masses: $H_2$, 2.016 g; He, 4.003 g; 0°C = 273 K

$$2.27 \text{ g } H_2 \times \frac{1 \text{ mol } H_2}{2.016 \text{ g } H_2} = 1.126 \text{ mol } H_2$$

$$1.03 \text{ g He} \times \frac{1 \text{ mol He}}{4.003 \text{ g He}} = 0.2573 \text{ mol He}$$

$$P_{H_2} = \frac{nRT}{V} = \frac{(1.126 \text{ mol } H_2)(0.08206 \text{ L-atm/mol-K})(273 \text{ K})}{(5.00 \text{ L})} = 5.05 \text{ atm}$$

$$P_{He} = \frac{nRT}{V} = \frac{(0.2573 \text{ mol He})(0.08206 \text{ L-atm/mol-K})(273 \text{ K})}{(5.00 \text{ L})} = 1.15 \text{ atm}$$

c. molar mass of Ar = 39.95 g; 27°C = 300 K

$$42.5 \text{ g Ar} \times \frac{1 \text{ mol Ar}}{39.95 \text{ g Ar}} = 1.064 \text{ mol Ar}$$

$$P = \frac{nRT}{V} = \frac{(1.064 \text{ mol Ar})(0.08206 \text{ L-atm/mol-K})(300 \text{ K})}{(9.97 \text{ L})} = 2.63 \text{ atm}$$

32. molar masses: $CaCO_3$, 100.09 g; $CO_2$, 44.01 g

$$1.25 \text{ g } CaCO_3 \times \frac{1 \text{ mol } CaCO_3}{100.09 \text{ g}} = 0.01249 \text{ mol } CaCO_3$$

$$0.01249 \text{ mol } CaCO_3 \times \frac{1 \text{ mol } CO_2}{1 \text{ mol } CaCO_3} = 0.01249 \text{ mol } CO_2$$

$$0.01249 \text{ mol } CO_2 \times \frac{44.01 \text{ g } CO_2}{1 \text{ mol } CO_2} = 0.550 \text{ g } CO_2$$

$$0.01249 \text{ mol } CO_2 \times \frac{22.4 \text{ L}}{1 \text{ mol}} = 0.280 \text{ L } CO_2 \text{ at STP}$$

34. a. mass of solution = 2.05 g NaCl + 19.2 g water = 21.25 g solution

$$\frac{2.05 \text{ g NaCl}}{21.25 \text{ g solution}} \times 100 = 9.65\% \text{ NaCl}$$

Review: Chapters 13, 14, and 15

b. $26.2 \text{ g solution} \times \dfrac{10.5 \text{ g CaCl}_2}{100 \text{ g solution}} = 2.75 \text{ g CaCl}_2$

c. $225 \text{ g solution} \times \dfrac{5.05 \text{ g NaCl}}{100 \text{ g solution}} = 11.4 \text{ g NaCl required}$

36. $M_1 \times V_1 = M_2 \times V_2$

   a. $M_2 = \dfrac{(12.5 \text{ mL})(1.515 \; M)}{(12.5 + 25 \text{ mL})} = 0.505 \; (0.51) \; M$

   b. $M_2 = \dfrac{(75.0 \text{ mL})(0.252 \; M)}{(225 \text{ mL})} = 0.0840 \; M$

   c. $M_2 = \dfrac{(52.1 \text{ mL})(0.751 \; M)}{(52.1 + 250. \text{ mL})} = 0.130 \; M$

38. a. $125 \text{ mL solution} \times \dfrac{1.84 \text{ g solution}}{1 \text{ mL solution}} = 230. \text{ g solution}$

   $230. \text{ g solution} \times \dfrac{98.3 \text{ g H}_2\text{SO}_4}{1 \text{ g solution}} = 226 \text{ g H}_2\text{SO}_4$

   b. The concentrated solution contains 226 g of H$_2$SO$_4$ (molar mass 98.09 g) in 125 mL (0.125 L) of solution

   $226 \text{ g H}_2\text{SO}_4 \times \dfrac{1 \text{ mol H}_2\text{SO}_4}{98.09 \text{ g H}_2\text{SO}_4} = 2.304 \text{ mol H}_2\text{SO}_4$

   $M = \dfrac{2.304 \text{ mol H}_2\text{SO}_4}{0.125 \text{ L solution}} = 18.4 \; M$

   c. $M_1 \times V_1 = M_2 \times V_2$

   $M_2 = \dfrac{(0.125 \text{ L})(18.4 \; M)}{3.01 \text{ L}} = 0.764 \; M$

   d. $\dfrac{0.764 \text{ mol}}{1 \text{ L}} \times \dfrac{2 \text{ equivalents}}{1 \text{ mol}} = 1.53 \; N$

   e. $\text{mmol NaOH} = 45.3 \text{ mL} \times \dfrac{0.532 \text{ mmol NaOH}}{1 \text{ mL}} = 24.10 \text{ mmol}$

   H$_2$SO$_4$ + 2NaOH → Na$_2$SO$_4$ + 2H$_2$O

   $\text{mmol H}_2\text{SO}_4 \text{ required} = 24.10 \text{ mmol NaOH} \times \dfrac{1 \text{ mmol H}_2\text{SO}_4}{2 \text{ mmol NaOH}} = 12.05 \text{ mmol H}_2\text{SO}_4$

   $12.05 \text{ mmol H}_2\text{SO}_4 \times \dfrac{1 \text{ mL solution}}{0.764 \text{ mmol H}_2\text{SO}_4} = 15.8 \text{ mL of the sulfuric acid solution.}$

Review: Chapters 13, 14, and 15

40. First, determine the total mass of reactants.

$20.0 \text{ mL} \times \dfrac{1.103 \text{ g}}{\text{mL}} = 22.1 \text{ g}$   Total mass of reactants: 22.1 g + 13.5 g = 35.6 g

Next, determine the mass of gas produced.   $1.473 \text{ L} \times \dfrac{1.798 \text{ g}}{\text{L}} = 2.648 \text{ g of gas}$

The law of conservation of mass states that the total mass of products must equal the total mass of reactants used (35.6 g). Therefore, the mass of the mixture remaining in the flask is:

35.6 g – 2.648 g = 32.9 g of mixture remaining in the flask after the reaction takes place

42. a. Using the solubility rules (Table 7.1), write out each reaction to determine the products and whether a precipitate forms.

|  | $CaCl_2$ | $Pb(NO_3)_2$ | $(NH_4)_3PO_4$ |
|---|---|---|---|
| $Na_2CO_3$ | $CaCO_3$ | $PbCO_3$ | No |
| $AgNO_3$ | $AgCl$ | No | $Ag_3PO_4$ |
| $K_2SO_4$ | $CaSO_4$ | $PbSO_4$ | No |

b. $M_1 \times V_1 = M_2 \times V_2$

$M_1 = 0.250 \text{ M}$        $M_2 = 2.00 \text{ M}$

$V_1 = 300.0 \text{ mL} = 0.3000 \text{ L}$        $V_2 = ?$

$V_2 = \dfrac{(0.250 \text{ M})(0.3000 \text{ L})}{(2.00 \text{ M})} = 0.0375 \text{ L} = 37.5 \text{ mL}$

Measure out 37.5 mL of 2.00 M $CaCl_2$, place it into an empty flask, and add water until the 300.0 mL line is reached.

# CHAPTER 16

# Acids and Bases

2. $HCl(g) \xrightarrow{H_2O} H^+(aq) + Cl^-(aq)$

   $NaOH(s) \xrightarrow{H_2O} Na^+(aq) + OH^-(aq)$

4. Conjugate acid–base pairs differ from each other by one proton (one hydrogen ion, $H^+$). For example, $CH_3COOH$ (acetic acid), differs from its conjugate base, $CH_3COO^-$ (acetate ion), by a single $H^+$ ion.

   $CH_3COOH(aq) \rightleftharpoons CH_3COO^-(aq) + H^+(aq)$

6. acids; bases

8. a. not a conjugate pair

   $H_2SO_4$, $HSO_4^-$

   $HSO_4^-$, $SO_4^{2-}$

   b. a conjugate pair: the two species differ by one proton

   c. not a conjugate pair

   $HClO_4$, $ClO_4^-$

   $HCl$, $Cl^-$

   d. not a conjugate pair

   $NH_4^+$, $NH_3$

   $NH_3$, $NH_2^-$

10. a. $NH_3$ (base) + $H_2O$ (acid) $\rightleftharpoons$ $NH_4^+$ (acid) + $OH^-$ (base)

    b. $PO_4^{3-}$ (base) + $H_2O$ (acid) $\rightleftharpoons$ $HPO_4^{2-}$ (acid) + $OH^-$ (base)

    c. $C_2H_3O_2^-$ (base) + $H_2O$ (acid) $\rightleftharpoons$ $HC_2H_3O_2$ (acid) + $OH^-$ (base)

12. The conjugate *acid* of the species indicated would have *one additional proton*:

    a. HClO

    b. HCl

    c. $HClO_3$

    d. $HClO_4$

# Chapter 16: Acids and Bases

14. The conjugate *bases* of the species indicated would have *one less proton*:
    a. BrO$^-$
    b. NO$_2^-$
    c. SO$_3^{2-}$
    d. CH$_3$NH$_2$

16. a. $O^{2-}(aq) + H_2O(l) \rightleftharpoons OH^-(aq) + OH^-(aq)$
    b. $NH_3(aq) + H_2O(l) \rightleftharpoons NH_4^+(aq) + OH^-(aq)$
    c. $HSO_4^-(aq) + H_2O(l) \rightleftharpoons SO_4^{2-}(aq) + H_3O^+(aq)$
    d. $HNO_2(aq) + H_2O(l) \rightleftharpoons NO_2^-(aq) + H_3O^+(aq)$

18. To say that an acid is *weak* in aqueous solution means that the acid does not easily transfer protons to water (and does not fully ionize). If an acid does not lose protons easily, then the acid's anion must be a strong attractor of protons (good at holding on to protons).

20. A strong acid is one that loses its protons easily and fully ionizes in water; this means that the acid's conjugate base must be poor at attracting and holding on to protons, and is therefore a relatively weak base. A weak acid is one that resists loss of its protons and does not ionize well in water; this means that the acid's conjugate base attracts and holds onto protons tightly and is a relatively strong base.

22. H$_2$SO$_4$ (sulfuric): $H_2SO_4 + H_2O \rightarrow HSO_4^- + H_3O^+$

    HCl (hydrochloric): $HCl + H_2O \rightarrow Cl^- + H_3O^+$

    HNO$_3$ (nitric): $HNO_3 + H_2O \rightarrow NO_3^- + H_3O^+$

    HClO$_4$ (perchloric): $HClO_4 + H_2O \rightarrow ClO_4^- + H_3O^+$

24. An oxyacid is an acid containing a particular element which is bonded to one or more oxygen atoms. HNO$_3$, H$_2$SO$_4$, HClO$_4$ are oxyacids. HCl, HF, HBr are not oxyacids.

26. Salicylic acid is a monoprotic acid: only the hydrogen of the carboxyl group ionizes.

    [Structural equation: 2-hydroxybenzoic acid (–COOH, –OH on benzene ring) + H$_2$O ⇌ 2-hydroxybenzoate (–COO$^-$, –OH on benzene ring) + H$_3$O$^+$]

28. For example, HCO$_3^-$ can behave as an acid if it reacts with something that more strongly gains protons than does HCO$_3^-$ itself. For example, HCO$_3^-$ would behave as an acid when reacting with hydroxide ion (a much stronger base).

    $HCO_3^-(aq) + OH^-(aq) \rightarrow CO_3^{2-}(aq) + H_2O(l)$.

    On the other hand, HCO$_3^-$ would behave as a base when reacted with something that more readily loses protons than does HCO$_3^-$ itself. For example, HCO$_3^-$ would behave as a base when reacting with hydrochloric acid (a much stronger acid).

## Chapter 16: Acids and Bases

$HCO_3^-(aq) + HCl(aq) \rightarrow H_2CO_3(aq) + Cl^-(aq)$

For $H_2PO_4^-$, similar equations can be written:

$H_2PO_4^-(aq) + OH^-(aq) \rightarrow HPO_4^{2-}(aq) + H_2O(l)$

$H_2PO_4^-(aq) + H_3O^+(aq) \rightarrow H_3PO_4(aq) + H_2O(l)$

30. The hydrogen ion concentration and the hydroxide ion concentration of water are *not* independent: they are related by the equilibrium

$H_2O(l) \rightleftharpoons H^+(aq) + OH^-(aq)$

for which $K_w = [H^+][OH^-] = 1.0 \times 10^{-14}$ at 25°C.

If the concentration of one of these ions is increased by addition of a reagent producing $H^+$ or $OH^-$, then the concentration of the complementary ion will have to decrease so that the value of $K_w$ will hold true. So if an acid is added to a solution, the concentration of hydroxide ion in the solution will decrease to a lower value. Similarly, if a base is added to a solution, then the concentration of hydrogen ion will have to decrease to a lower value.

32. $K_w = [H^+][OH^-] = 1.0 \times 10^{-14}$ at 25°C

    a. $[H^+] = \dfrac{1.0 \times 10^{-14}}{7.86 \times 10^{-4} \, M} = 1.3 \times 10^{-11} \, M$; solution is basic

    b. $[H^+] = \dfrac{1.0 \times 10^{-14}}{5.44 \times 10^{-8} \, M} = 1.8 \times 10^{-7} \, M$; solution is acidic

    c. $[H^+] = \dfrac{1.0 \times 10^{-14}}{3.19 \times 10^{-3} \, M} = 3.1 \times 10^{-12} \, M$; solution is basic

    d. $[H^+] = \dfrac{1.0 \times 10^{-14}}{2.51 \times 10^{-9} \, M} = 4.0 \times 10^{-6} \, M$; solution is acidic

34. $K_w = [H^+][OH^-] = 1.0 \times 10^{-14}$ at 25°C

    a. $[OH^-] = \dfrac{1.0 \times 10^{-14}}{1.02 \times 10^{-7} \, M} = 9.8 \times 10^{-8} \, M$; solution is acidic

    b. $[OH^-] = \dfrac{1.0 \times 10^{-14}}{9.77 \times 10^{-8} \, M} = 1.02 \times 10^{-7} \, M \, (1.0 \times 10^{-7} \, M)$; solution is slightly basic

    c. $[OH^-] = \dfrac{1.0 \times 10^{-14}}{3.41 \times 10^{-3} \, M} = 2.9 \times 10^{-12} \, M$; solution is acidic

    d. $[OH^-] = \dfrac{1.0 \times 10^{-14}}{4.79 \times 10^{-11} \, M} = 2.1 \times 10^{-4} \, M$; solution is basic

36.   a. $[OH^-] = 5.05 \times 10^{-5} \, M$ is more basic

    b. $[OH^-] = 4.21 \times 10^{-6} \, M$ is more basic

    c. $[H^+] = 1.25 \times 10^{-12} \, M$ is more basic

38. Answer depends on student choice.

Chapter 16: Acids and Bases

40. pH 1–2, deep red; pH 4, purple; pH 8, blue; pH 11, green

42. pH = –log[H$^+$]
   a. pH = –log[0.00512 M] = 2.291; solution is acidic
   b. pH = –log[3.76 × 10$^{-5}$ M] = 4.425; solution is acidic
   c. pH = –log[5.61 × 10$^{-10}$ M] = 9.251; solution is basic
   d. pH = –log[8.44 × 10$^{-6}$ M] = 5.074; solution is acidic

44. pOH = –log[OH$^-$]    pH = 14.00 – pOH
   a. pOH = –log[4.85 × 10$^{-5}$ M] = 4.314
      pH = 14.00 – 4.314 = 9.686 = 9.69; solution is basic
   b. pOH = –log[3.96 × 10$^{-7}$ M] = 6.402
      pH = 14.00 – 6.402 = 7.598 = 7.60; solution is basic
   c. pOH = –log[1.22 × 10$^{-10}$ M] = 9.914
      pH = 14.00 – 9.914 = 4.086 = 4.09; solution is acidic
   d. pOH = –log[5.33 × 10$^{-12}$ M] = 11.273
      pH = 14.00 – 11.273 = 2.727 = 2.73; solution is acidic

46. pOH = 14.00 – pH
   a. pOH = 14.00 – 10.75 = 3.25; solution is basic
   b. pOH = 14.00 – 3.66 = 10.34; solution is acidic
   c. pOH = 14.00 – 1.98 = 12.02; solution is acidic
   d. pOH = 14.00 – 12.47 = 1.53; solution is basic

48. 
   a. pH = –log[1.91 × 10$^{-2}$ M] = 1.719; solution is acidic
      $$[OH^-] = \frac{1.0 \times 10^{-14}}{1.91 \times 10^{-2} \ M} = 5.2 \times 10^{-13} \ M$$
   b. pH = –log[4.83 × 10$^{-7}$ M] = 6.316; solution is acidic
      $$[OH^-] = \frac{1.0 \times 10^{-14}}{4.83 \times 10^{-7} \ M} = 2.1 \times 10^{-8} \ M$$
   c. pH = –log[8.92 × 10$^{-11}$ M] = 10.050; solution is basic
      $$[OH^-] = \frac{1.0 \times 10^{-14}}{8.92 \times 10^{-11} \ M} = 1.1 \times 10^{-4} \ M$$
   d. pH = –log[6.14 × 10$^{-5}$ M] = 4.212; solution is acidic
      $$[OH^-] = \frac{1.0 \times 10^{-14}}{6.14 \times 10^{-5} \ M} = 1.6 \times 10^{-10} \ M$$

Chapter 16:   Acids and Bases

50. $[H^+] = \{inv\}\{log\}[-pH]$ or $10^{-pH}$
   a. $[H^+] = \{inv\}\{log\}[-2.75] = 0.0018\ M$
   b. $[H^+] = \{inv\}\{log\}[-12.8] = 2 \times 10^{-13}\ M$
   c. $[H^+] = \{inv\}\{log\}[-4.33] = 4.7 \times 10^{-5}\ M$
   d. $[H^+] = \{inv\}\{log\}[-9.61] = 2.5 \times 10^{-10}\ M$

52. $pH + pOH = 14.00$       $[H^+] = \{inv\}\{log\}[-pH]$ or $10^{-pH}$
   a. $pH = 14.00 - 4.99 = 9.01$
      $[H^+] = \{inv\}\{log\}[-9.01] = 9.8 \times 10^{-10}\ M$
   b. $[H^+] = \{inv\}\{log\}[-7.74] = 1.8 \times 10^{-8}\ M$
   c. $pH = 14.00 - 10.74 = 3.26$
      $[H^+] = \{inv\}\{log\}[-3.26] = 5.5 \times 10^{-4}\ M$
   d. $[H^+] = \{inv\}\{log\}[-2.25] = 5.6 \times 10^{-3}\ M$

54. a. $pH = -\log[3.42 \times 10^{-10}\ M] = 9.466$
   b. $pH = 14.00 - pOH = 14.00 - 5.92 = 8.08$
   c. $pOH = -\log[2.86 \times 10^{-7}\ M] = 6.544$     $pH = 14.00 - 6.544 = 7.46$
   d. $pH = -\log[9.11 \times 10^{-2}\ M] = 1.040$

56. The solution contains water molecules, $H_3O^+$ ions (protons), and $NO_3^-$ ions. Because $HNO_3$ is a strong acid, which is completely ionized in water, there are no $HNO_3$ molecules present.

58. a. $HNO_3$ is a strong acid and completely ionized so $[H^+] = 1.21 \times 10^{-3}\ M$ and $pH = 2.917$.
   b. $HClO_4$ is a strong acid and completely ionized so $[H^+] = 0.000199\ M$ and $pH = 3.701$.
   c. $HCl$ is a strong acid and completely ionized so $[H^+] = 5.01 \times 10^{-5}\ M$ and $pH = 4.300$.
   d. $HBr$ is a strong acid and completely ionized so $[H^+] = 0.00104\ M$ and $pH = 2.983$.

60. A buffered solution consists of a mixture of a weak acid and its conjugate base; one example of a buffered solution is a mixture of acetic acid ($CH_3COOH$) and sodium acetate ($NaCH_3COO$).

62. The weak acid component of a buffered solution is capable of reacting with added strong base. For example, using the buffered solution given as an example in Question 60, acetic acid would consume added sodium hydroxide as follows:

   $CH_3COOH(aq) + NaOH(aq) \rightarrow NaCH_3COO(aq) + H_2O(l)$.

   Acetic acid *neutralizes* the added NaOH and prevents it from having much effect on the overall pH of the solution.

64. HCl:    $H_3O^+ + C_2H_3O_2^- \rightarrow HC_2H_3O_2 + H_2O$
    NaOH:  $OH^- + HC_2H_3O_2 \rightarrow C_2H_3O_2^- + H_2O$

Chapter 16: Acids and Bases

66. a. NaOH is completely ionized, so [OH⁻] = 0.10 M.

   pOH = –log[0.10] = 1.00

   pH = 14.00 – 1.00 = 13.00

   b. KOH is completely ionized, so [OH⁻] = 2.0 × 10⁻⁴ M.

   pOH = –log[2.0 × 10⁻⁴] = 3.70

   pH = 14.00 – 3.70 = 10.30

   c. CsOH is completely ionized, so [OH⁻] = 6.2 × 10⁻³ M.

   pOH = –log[6.2 × 10⁻³] = 2.21

   pH = 14.00 – 2.21 = 11.79

   d. NaOH is completely ionized, so [OH⁻] = 0.0001 M.

   pOH = –log[0.0001] = 4.0

   pH = 14.00 – 4.0 = 10.0

68. (d)

70. a. $NO_2^-$ is a relatively strong base.

   b. $HCOO^-$ is a relatively strong base.

   c. $ClO_4^-$ is a very weak base (conjugate of a strong acid).

   d. $NO_3^-$ is a very weak base (conjugate of a strong acid).

72. Ordinarily in calculating the pH of strong acid solutions, the major contribution to the concentration of hydrogen ion present is from the dissolved strong acid; we ordinarily neglect the small amount of hydrogen ion present in such solutions due to the ionization of water. With 1.0 × 10⁻⁷ M HCl solution, however, the amount of hydrogen ion present due to the ionization of *water* is *comparable* to that present due to the addition of *acid* (HCl) and must be considered in the calculation of pH.

74. accepts

76. base

78. 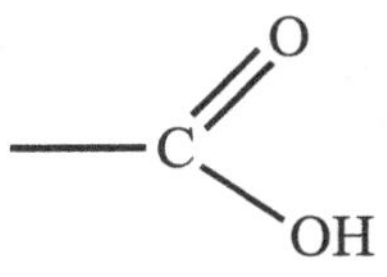 —C(=O)—OH     $CH_3COOH + H_2O \rightleftharpoons C_2H_3O_2^- + H_3O^+$

80. 1.0 × 10⁻¹⁴

82. lower

84. pH

86. weak acid

## Chapter 16: Acids and Bases

88. a. Equation 1:

   (acid1) + (base1) → (conjugate acid1) + (conjugate base1)

   Equation 2:

   (base2) + (acid2) → (conjugate acid2) + (conjugate base2)

   The acids are the proton donors and the bases are the proton acceptors. By looking at which species is positively charged or negatively charged in the products, it's possible to determine which reactant is the proton donor and which is the proton acceptor.

   b. An Arrhenius acid produces hydrogen ions. An Arrhenius base produces hydroxide ions. Therefore, acid1 is considered an Arrhenius acid. A Brønsted-Lowry acid is a proton donor and a Brønsted-Lowry base is a proton acceptor. Thus, acid1 and acid2 are both Brønsted-Lowry acids and base1 and base2 are both Brønsted-Lowry bases.

90. The conjugate *acid* of the species indicated would have *one additional proton*:

   a. $NH_4^+$
   b. $NH_3$
   c. $H_3O^+$
   d. $H_2O$

92. a. A buffer: $HClO_2$ and $KClO_2$ are conjugates.
   b. Not a buffer: $S^{2-}$ (of $Na_2S$) is not the conjugate base of $H_2S$.
   c. Not a buffer: NaHCO is not the conjugate base of HCOOH.
   d. Not a buffer: $HClO_4$ is not the conjugate base of HClO.

94. $K_w = [H^+][OH^-] = 1.0 \times 10^{-14}$ at 25°C

   a. $[H^+] = \dfrac{1.0 \times 10^{-14}}{4.22 \times 10^{-3} \, M} = 2.4 \times 10^{-12} \, M$; solution is basic

   b. $[H^+] = \dfrac{1.0 \times 10^{-14}}{1.01 \times 10^{-13} \, M} = 9.9 \times 10^{-2} \, M$; solution is acidic

   c. $[H^+] = \dfrac{1.0 \times 10^{-14}}{3.05 \times 10^{-7} \, M} = 3.3 \times 10^{-8} \, M$; solution is basic

   d. $[H^+] = \dfrac{1.0 \times 10^{-14}}{6.02 \times 10^{-6} \, M} = 1.7 \times 10^{-9} \, M$; solution is basic

96. pH + pOH = 14.00

   a. pH = 14.00 – 4.32 = 9.68; solution is basic
   b. pH = 14.00 – 8.90 = 5.10; solution is acidic
   c. pH = 14.00 – 1.81 = 12.19; solution is basic
   d. pH = 14.00 – 13.1 = 0.9; solution is acidic

98. $pOH = -\log[OH^-]$    $pH = 14.00 - pOH$

   a. $pOH = -\log[1.4 \times 10^{-6} M] = 5.85$; $pH = 14.00 - 5.85 = 8.15$; solution is basic
   b. $pOH = -\log[9.35 \times 10^{-9} M] = 8.029 = 8.03$; $pH = 14.00 - 8.029 = 5.97$; solution is acidic
   c. $pOH = -\log[2.21 \times 10^{-1} M] = 0.656 = 0.66$; $pH = 14.00 - 0.656 = 13.34$; solution is basic
   d. $pOH = -\log[7.98 \times 10^{-12} M] = 11.10$; $pH = 14.00 - 11.098 = 2.90$; solution is acidic

100. a. $[OH^-] = \dfrac{1.0 \times 10^{-14}}{5.72 \times 10^{-4} \, M} = 1.75 \times 10^{-11} \, M = 1.8 \times 10^{-11} \, M$

   $pOH = -\log[1.75 \times 10^{-11} M] = 10.76$

   $pH = 14.00 - 10.76 = 3.24$

   b. $[H^+] = \dfrac{1.0 \times 10^{-14}}{8.91 \times 10^{-5} \, M} = 1.12 \times 10^{-10} \, M = 1.1 \times 10^{-10} \, M$

   $pH = -\log[1.12 \times 10^{-10} M] = 9.95$

   $pOH = 14.00 - 9.95 = 4.05$

   c. $[OH^-] = \dfrac{1.0 \times 10^{-14}}{2.87 \times 10^{-12} \, M} = 3.48 \times 10^{-3} \, M = 3.5 \times 10^{-3} \, M$

   $pOH = -\log[3.48 \times 10^{-3} M] = 2.46$

   $pH = 14.00 - 2.46 = 11.54$

   d. $[H^+] = \dfrac{1.0 \times 10^{-14}}{7.22 \times 10^{-8} \, M} = 1.39 \times 10^{-7} \, M = 1.4 \times 10^{-7} \, M$

   $pH = -\log[1.39\times \times 10^{-7} M] = 6.86$

   $pOH = 14.00 - 6.86 = 7.14$

102. $pH = 14.00 - pOH$        $[H^+] = \{inv\}\{log\}[-pH]$ or $10^{-pH}$

   a. $[H^+] = \{inv\}\{log\}[-5.41] = 3.9 \times 10^{-6} \, M$
   b. $pH = 14.00 - 12.04 = 1.96$    $[H^+] = \{inv\}\{log\}[-1.96] = 1.1 \times 10^{-2} \, M$
   c. $[H^+] = \{inv\}\{log\}[-11.91] = 1.2 \times 10^{-12} \, M$
   d. $pH = 14.00 - 3.89 = 10.11$    $[H^+] = \{inv\}\{log\}[-10.11] = 7.8 \times 10^{-11} \, M$

104. a. $HClO_4$ is a strong acid and completely ionized so $[H^+] = 1.4 \times 10^{-3} \, M$ and $pH = 2.85$.
   b. $HCl$ is a strong acid and completely ionized so $[H^+] = 3.0 \times 10^{-5} \, M$ and $pH = 4.52$.
   c. $HNO_3$ is a strong acid and completely ionized so $[H^+] = 5.0 \times 10^{-2} \, M$ and $pH = 1.30$.
   d. $HCl$ is a strong acid and completely ionized so $[H^+] = 0.0010 \, M$ and $pH = 3.00$.

106. a and d; The conjugate base has one less proton ($H^+$) compared to its acid counterpart.

# Chapter 16: Acids and Bases

108. NaCl: neutral ($Na^+$ and $Cl^-$ are very weak conjugates.)
RbOCl: basic ($Rb^+$ is a very weak conjugate. $OCl^-$ is a weak base.)
KI: neutral ($K^+$ and $I^-$ are very weak conjugates.)
$Ba(ClO_4)_2$: neutral ($Ba^{2+}$ and $ClO_4^-$ are very weak conjugates.)
$NH_4NO_3$: acidic ($NH_4^+$ is a weak acid. $NO_3^-$ is a very weak conjugate.)

# CHAPTER 17

# Equilibrium

2. Two H–H bonds in the two $H_2$ molecules and one O=O bond in the $O_2$ molecule must be broken. Four O–H bonds in the two $H_2O$ molecules must form.

4. activation energy

6. Enzymes are biochemical catalysts that speed up the complicated reactions that would be too slow to sustain life at normal body temperatures. Carbonic anhydrase speeds up the reaction between carbon dioxide and water to help prevent an excess accumulation of carbon dioxide in our blood.

8. The equilibrium state occurs when the rate of evaporation exactly equals the rate of condensation. The vapor pressure and liquid level remain constant because exactly the same number of molecules escapes the liquid as return to it.

10. Chemical equilibrium occurs when two *opposing* chemical reactions reach the *same speed* in a closed system. When a state of chemical equilibrium has been reached, the concentrations of reactants and products present in the system remain *constant* with time, and the reaction appears to "stop." A chemical reaction that reaches a state of equilibrium is indicated by using a double arrow ($\rightleftharpoons$). The points of the double arrow point in opposite directions, to indicate that two opposite processes are going on.

12. a. The green line is $H_2$ because hydrogen is initially present in the greatest concentration. The blue line is $N_2$ because some nitrogen is initially present but not as much as the $H_2$ (a third of the amount). The pink line is $NH_3$ because at first no product is present but then $N_2$ and $H_2$ react to form $NH_3$.

    b. The concentrations of both $N_2$ and $H_2$ decrease at first because they react to form $NH_3$ (which then causes the concentration of $NH_3$ to go up). None of the concentrations become zero over time because eventually some of the $NH_3$ shifts back to form $N_2$ and $H_2$ again. Eventually the concentration of each species remains constant because the rate of the forward reaction equals the rate of the backward reaction (equilibrium is reached).

    c. Equilibrium is reached when the lines become straight (the concentration over time does not change). As stated in *b*, the rate of the forward reaction equals the rate of the backward reaction.

14. The equilibrium constant is a *ratio* of concentration of products to concentration of reactants, with all concentrations measured at equilibrium. Depending on the amount of reactant present at the beginning of an experiment, there may be different absolute amounts of reactants and products present at equilibrium, but the *ratio* will always be the same for a given reaction at a given temperature. For example, the ratios (4/2) and (6/3) are different absolutely in terms of the numbers involved, but each of these ratios has the *value* of 2.

Chapter 17:  Equilibrium

16. a.  $K = \dfrac{[NCl_3(g)]^2}{[N_2(g)][Cl_2(g)]^3}$

    b.  $K = \dfrac{[HI(g)]^2}{[H_2(g)][I_2(g)]}$

    c.  $K = \dfrac{[N_2H_4(g)]}{[N_2(g)][H_2(g)]^2}$

18. a.  $K = \dfrac{[CH_3OH]}{[CO][H_2]^2}$

    b.  $K = \dfrac{[NO]^2[O_2]}{[NO_2]^2}$

    c.  $K = \dfrac{[PBr_3]^4}{[P_4][Br_2]^6}$

20. $N_2(g) + 3H_2(g) \rightleftharpoons 2NH_3(g)$

    $K = \dfrac{[NH_3(g)]^2}{[N_2(g)][H_2(g)]^3} = \dfrac{[0.25\ M]^2}{[5.3 \times 10^{-5}\ M][3.4 \times 10^{-3}\ M]^3} = 3.0 \times 10^{10}$

22. $2N_2O(g) + O_2(g) \rightleftharpoons 4NO(g)$

    $K = \dfrac{[NO]^4}{[N_2O]^2[O_2]} = \dfrac{[0.00341\ M]^4}{[0.0293\ M]^2[0.0325\ M]} = 4.85 \times 10^{-6}$

24. True. Equilibrium constants represent ratios of the *concentrations* of products and reactants present at the point of equilibrium. The *concentration* of a pure solid or of a pure liquid is constant and is determined by the density of the solid or liquid. For example, suppose you had a liter of water. Within that liter of water are 55.5 mol of water (the number of moles of water that is contained in one liter of water *does not vary*).

26. a.  $K = [H_2O(g)][CO_2(g)]$

    b.  $K = [CO_2(g)]$

    c.  $K = \dfrac{1}{[O_2(g)]^3}$

28. a.  incorrect; $K = [N_2(g)][Br_2(g)]^3$; solids are not included in the $K$ expression

    b.  incorrect; $K = [H_2O(g)]/[H_2(g)]$; both solids and liquids are not included in the $K$ expression

    c.  correct

Chapter 17:   Equilibrium

30. When an additional amount of product is added to an equilibrium system, the system shifts to the left and adjusts so as to increase some of the added reactants. This results in a net *increase* in the amount of reactants, compared to the equilibrium system before the additional product was added, and so the amount of $O_2(g)$ in the system will be higher than if the additional $CO_2(g)$ had not been added. The numerical *value* of the equilibrium constant does *not* change when a product is added: the concentrations of all reactants and products adjust until the correct value of $K$ is once again achieved.

32. If heat is applied to an exothermic reaction (the temperature is raised), the equilibrium shifts to the left. Less product will be present at equilibrium than if the temperature had not been increased. The value of $K$ decreases because the amount of product decreases ($K$ is a special ratio of the concentrations of the products to the concentrations of the reactants).

34. a. shifts left (system reacts to the left to replace fluorine as it is removed)
    b. no change (phosphorus is in the *solid* state)
    c. shifts right (system reacts to the right to replace $PF_3$ as it is removed)

36. a. no change (B is solid)
    b. shift right (system reacts to replace removed C)
    c. shift left (system reacts by shifting in direction of fewer mol of gas)
    d. shift right (the reaction is endothermic as written)

38. The answer is d. When hydrogen gas is added, equilibrium will shift away from the addition of reactant and toward the product side, producing more water vapor. The value of $K$ does not change.

40. For an endothermic reaction, an increase in temperature will shift the position of equilibrium to the right (toward the products).

42. $CO(g) + 2H_2(g) \rightleftharpoons CH_3OH(l)$

    remove some $CO(g)$ or $H_2(g)$: the system will react in the backward direction to replace $CO(g)$ or $H_2(g)$ as it is removed
    increase the volume of the system: the system will react in the direction to create more moles of gas
    increase the temperature: addition of heat favors reactants in an exothermic reaction

44. A small equilibrium constant implies that not much product forms before equilibrium is reached. The reaction would not be a good source of the products unless Le Châtelier's principle can be used to force the reaction to the right.

46. $K = \dfrac{[SO_3][NO]}{[SO_2][NO_2]} = \dfrac{[4.99 \times 10^{-5} \, M][6.31 \times 10^{-7} \, M]}{[2.11 \times 10^{-2} \, M][1.73 \times 10^{-3} \, M]} = 8.63 \times 10^{-7}$

48. $K = 5.21 \times 10^{-3} = \dfrac{[CO][H_2O]}{[CO_2][H_2]} = \dfrac{[4.73 \times 10^{-3} \, M][5.21 \times 10^{-3} \, M]}{[3.99 \times 10^{-2} \, M][H_2]}$

    $[H_2] = 0.119 \, M$

Chapter 17:  Equilibrium

50.  $K = 2.4 \times 10^{-3} = \dfrac{[H_2]^2[O_2]}{[H_2O]^2} = \dfrac{[1.9 \times 10^{-2}]^2[O_2]}{[1.1 \times 10^{-1}]^2}$

$[O_2] = 8.0 \times 10^{-2}\ M$

52.  $K = 8.1 \times 10^{-3} = \dfrac{[NO_2]^2}{[N_2O_4]} = \dfrac{[NO_2]^2}{[5.4 \times 10^{-4}\ M]}$

$[NO_2] = 2.1 \times 10^{-3}\ M$

54.  solubility product, $K_{sp}$

56.  Stirring or grinding the solute increases the speed with which the solute dissolves, but the ultimate *amount* of solute that dissolves is fixed by the equilibrium constant for the dissolving process, $K_{sp}$, which changes only with temperature. Therefore only the temperature will affect the solubility.

58.  
a.  $NiS(s) \rightleftharpoons Ni^{2+}(aq) + S^{2-}(aq)$   $K_{sp} = [Ni^{2+}(aq)][S^{2-}(aq)]$
b.  $CuCO_3(s) \rightleftharpoons Cu^{2+}(aq) + CO_3^{2-}(aq)$   $K_{sp} = [Cu^{2+}(aq)][CO_3^{2-}(aq)]$
c.  $BaCrO_4(s) \rightleftharpoons Ba^{2+}(aq) + CrO_4^{2-}(aq)$   $K_{sp} = [Ba^{2+}(aq)][CrO_4^{2-}(aq)]$
d.  $Ag_3PO_4(s) \rightleftharpoons 3Ag^+(aq) + PO_4^{3-}(aq)$   $K_{sp} = [Ag^+(aq)]^3[PO_4^{3-}(aq)]$

60.  $MgCO_3(s) \rightleftharpoons Mg^{2+}(aq) + CO_3^{2-}(aq)$

Molar mass $MgCO_3 = 84.32$ g

Let $x$ represent the solubility of $MgCO_3$ in mol/L. Then $[CO_3^{2-}] = x$ and $[Mg^{2+}] = x$ from the stoichiometry of the equation.

$K_{sp} = [Mg^{2+}][CO_3^{2-}] = 3.5 \times 10^{-8} = (x)(x) = x^2$

then the molar solubility of $MgCO_3 = x = 1.87 \times 10^{-4}\ M\ (1.9 \times 10^{-4}\ M)$

$1.87 \times 10^{-4}\ \dfrac{mol}{L} \times \dfrac{84.32\ g}{1\ mol} = 0.016\ g/L$

62.  $Ni(OH)_2(s) \rightleftharpoons Ni^{2+}(aq) + 2OH^-(aq)$

molar mass $Ni(OH)_2 = 92.71$ g

let $x$ represent the molar solubility of $Ni(OH)_2$: then $[Ni^{2+}] = x$ and $[OH^-] = 2x$.

$K_{sp} = [Ni^{2+}][OH^-]^2 = 2.0 \times 10^{-15} = [x][2x]^2 = 4x^3$

then the molar solubility of $Ni(OH)_2 = x = 7.9 \times 10^{-6}\ M$

gram solubility = $7.98 \times 10^{-6}\ \dfrac{mol}{L} \times \dfrac{92.71\ g}{1\ mol} = 7.4 \times 10^{-4}\ g/L$

64.  $CaSO_4(s) \rightleftharpoons Ca^{2+}(aq) + SO_4^{2-}(aq)$

molar mass $CaSO_4 = 136.15$ g

$$2.05 \frac{g}{L} \times \frac{1 \text{ mol}}{136.15 \text{ g}} = 1.506 \times 10^{-2} \text{ M}$$

If $CaSO_4$ dissolves to the extent of $1.506 \times 10^{-2}$ M, then $[Ca^{2+}]$ will be $1.506 \times 10^{-2}$ M and $[SO_4^{2-}]$ will be $1.506 \times 10^{-2}$ M also.

$K_{sp} = [Ca^{2+}][SO_4^{2-}] = [1.506 \times 10^{-2} \text{ M}][1.506 \times 10^{-2} \text{ M}] = 2.27 \times 10^{-4}$

66.  $Cr(OH)_3(s) \rightleftharpoons Cr^{3+}(aq) + 3OH^-(aq)$

If $Cr(OH)_3$ dissolves to the extent of $8.21 \times 10^{-5}$ M, then $[Cr^{3+}]$ will be $8.21 \times 10^{-5}$ M and $[OH^-]$ will be $3(8.21 \times 10^{-5}$ M) in a saturated solution.

$K_{sp} = [Cr^{3+}][OH^-]^3 = [8.21 \times 10^{-5} \text{ M}][2.46 \times 10^{-4} \text{ M}]^3 = 1.23 \times 10^{-15}$

68.  $PbCl_2(s) \rightleftharpoons Pb^{2+}(aq) + 2Cl^-(aq)$

$K_{sp} = [Pb^{2+}][Cl^-]^2$

If $PbCl_2$ dissolves to the extent of $3.6 \times 10^{-2}$ M, then $[Pb^{2+}] = 3.6 \times 10^{-2}$ M and $[Cl^-] = 2 \times (3.6 \times 10^{-2}) = 7.2 \times 10^{-2}$ M.

$K_{sp} = (3.6 \times 10^{-2} \text{ M})(7.2 \times 10^{-2} \text{ M})^2 = 1.9 \times 10^{-4}$

molar mass $PbCl_2 = 278.1$ g

$$\frac{3.6 \times 10^{-2} \text{ mol}}{1 \text{ L}} \times \frac{278.1 \text{ g}}{1 \text{ mol}} = 10. \text{ g/L}$$

70.  $Fe(OH)_3(s) \rightleftharpoons Fe^{3+}(aq) + 3OH^-(aq)$

$K_{sp} = [Fe^{3+}][OH^-]^3 = 4 \times 10^{-38}$

Let $x$ represent the number of moles of $Fe(OH)_3$ that dissolve per liter; then $[Fe^{3+}] = x$.

The amount of hydroxide ion that would be produced by the dissolving of the $Fe(OH)_3$ would then be $3x$, but pure water itself contains hydroxide ion at the concentration of $1.0 \times 10^{-7}$ M (see Chapter 17). The total concentration of hydroxide ion is then $[OH^-] = (3x + 1.0 \times 10^{-7})$. As $x$ must be a very small number [because $Fe(OH)_3$ is not very soluble], we can save ourselves a lot of arithmetic if we use the approximation that

$(3x + 1.0 \times 10^{-7} \text{ M}) = 1.0 \times 10^{-7}$

$K_{sp} = [x][1.0 \times 10^{-7}]^3 = 4 \times 10^{-38}$

$x = 4 \times 10^{-17}$ M

molar mass $Fe(OH)_3 = 106.9$ g

$$\frac{4 \times 10^{-17} \text{ mol}}{1 \text{ L}} \times \frac{106.9 \text{ g}}{1 \text{ mol}} = 4 \times 10^{-15} \text{ g/L}$$

72.  An increase in temperature increases the fraction of molecules that possess sufficient energy for a collision to result in a reaction.

74.  catalyst

## Chapter 17: Equilibrium

76. constant

78. When we say that a chemical equilibrium is *dynamic*, we are recognizing the fact that even though the reaction has appeared macroscopically to have stopped, on a microscopic basis the forward and reverse reactions are still taking place, at the same speed.

80. heterogeneous

82. position

84. The balanced equation is $H_2O(g) + CO(g) \rightleftharpoons H_2(g) + CO_2(g)$. Initially, 8 $H_2O$ molecules are present and 6 CO molecules are present in the same 1.0-L container. The system reaches equilibrium by

|  | $H_2O(g)$ + | $CO(g)$ $\rightleftharpoons$ | $H_2(g)$ + | $CO_2(g)$ |
|---|---|---|---|---|
| Initial | 8 | 6 | 0 | 0 |
| Change | $-x$ | $-x$ | $+x$ | $+x$ |
| Equilibrium | $8-x$ | $6-x$ | $x$ | $x$ |

To determine the value of $x$, use the equilibrium expression

$$K = \frac{[H_2][CO_2]}{[H_2O][CO]} = \frac{(x)(x)}{(8-x)(6-x)} = 2.0$$

Use the quadratic equation to solve for $x$. $x = 24, 4$. The value of $x$ cannot be 24 or else a negative equilibrium concentration for the reactants would result. Therefore, $x$ must equal 4. The number of each type of molecule in the container at equilibrium is

$H_2O = 8 - x = 8 - 4 = 4$

$CO = 6 - x = 6 - 4 = 2$

$H_2 = x = 4$

$CO_2 = x = 4$

86. An equilibrium reaction may come to many *positions* of equilibrium, but at each possible position of equilibrium, the numerical value of the equilibrium constant is fulfilled. If different amounts of reactant are taken in different experiments, the *absolute amounts* of reactant and product present at the point of equilibrium reached will differ from one experiment to another, but the *ratio* that defines the equilibrium constant will be the same.

88. $PCl_5(g) \rightleftharpoons PCl_3(g) + Cl_2(g)$

$$K = \frac{[PCl_3][Cl_2]}{[PCl_5]} = 4.5 \times 10^{-3}$$

The concentration of $PCl_5$ is to be twice the concentration of $PCl_3$: $[PCl_5] = 2 \times [PCl_3]$

$$K = \frac{[PCl_3][Cl_2]}{2 \times [PCl_3]} = 4.5 \times 10^{-3}$$

$$K = \frac{[Cl_2]}{2} = 4.5 \times 10^{-3} \quad \text{and} \quad [Cl_2] = 9.0 \times 10^{-3} \, M$$

90. As all of the metal carbonates indicated have the metal ion in the +2 oxidation state, we can illustrate the calculations for a general metal carbonate, $MCO_3$:

$$MCO_3(s) \rightleftharpoons M^{2+}(aq) + CO_3^{2-}(aq) \qquad K_{sp} = [M^{2+}(aq)][CO_3^{2-}(aq)]$$

If we then let $x$ represent the number of moles of $MCO_3$ that dissolve per liter, then $[M^{2+}(aq)] = x$ and $[CO_3^{2-}(aq)] = x$ also because the reaction is of 1:1 stoichiometry. Therefore,

$K_{sp} = [M^{2+}(aq)][CO_3^{2-}(aq)] = x^2$ for each salt. Solving for $x$ gives the following results.

$[BaCO_3] = x = 7.1 \times 10^{-5}\ M$

$[CdCO_3] = x = 2.3 \times 10^{-6}\ M$

$[CaCO_3] = x = 5.3 \times 10^{-5}\ M$

$[CoCO_3] = x = 3.9 \times 10^{-7}\ M$

92. Although a small solubility product generally implies a small solubility, comparisons of solubility based directly on $K_{sp}$ values are only valid if the salts produce the same numbers of positive and negative ions per formula when they dissolve. For example, one can compare the solubilities of $AgCl(s)$ and $NiS(s)$ directly using $K_{sp}$, because each salt produces one positive and one negative ion per formula when dissolved. One could not directly compare $AgCl(s)$ with a salt such as $Ca_3(PO_4)_2$, however.

94. At higher temperatures, the average kinetic energy of the reactant molecules is larger. At higher temperatures, the probability that a collision between molecules will be energetic enough for reaction to take place is larger. On a molecular basis, a higher temperature means a given molecule will be moving faster.

96. a. $K = \dfrac{[HBr]^2}{[H_2][Br_2]}$

b. $K = \dfrac{[H_2S]^2}{[H_2]^2[S_2]}$

c. $K = \dfrac{[HCN]^2}{[H_2][C_2N_2]}$

98. $N_2(g) + 3Cl_2(g) \rightleftharpoons 2NCl_3(g)$

$$K = \dfrac{[NCl_3(g)]^2}{[N_2(g)][Cl_2(g)]^3} = \dfrac{[1.9 \times 10^{-1}\ M]^2}{[1.4 \times 10^{-3}\ M][4.3 \times 10^{-4}\ M]^3} = 3.2 \times 10^{11}$$

100. a. $K = \dfrac{1}{[O_2]^3}$

b. $K = \dfrac{1}{[NH_3][HCl]}$

c. $K = \dfrac{1}{[O_2]}$

## Chapter 17: Equilibrium

102. The second snapshot is the first to represent an equilibrium mixture because after this point, the concentrations of the reactant and products remain constant. 6 molecules of $A_2B$ reacted initially.

$$2A_2B(g) \rightleftharpoons 2A_2(g) + B_2(g)$$

| | | | |
|---|---|---|---|
| Initial | ? | 0 | 0 |
| Change | $-2x$ | $+2x$ | $+x$ |
| Equilibrium | $?-2x=2$ | $2x=4$ | $x=2$ |

Therefore ? = 6.

104. The reaction is *exo*thermic as written. An increase in temperature (addition of heat) will shift the reaction to the left (toward reactants).

106. $K = \dfrac{[NH_3]^2}{[N_2][H_2]^3} = 1.3 \times 10^{-2} = \dfrac{[NH_3]^2}{[0.1M][0.1M]^3}$

$[NH_3]^2 = 1.3 \times 10^{-6}$

$[NH_3] = 1.1 \times 10^{-3} \, M$

108. a. $Cu(OH)_2(s) \rightleftharpoons Cu^{2+}(aq) + 2OH^-(aq)$

$K_{sp} = [Cu^{2+}][OH^-]^2$

b. $Cr(OH)_3(s) \rightleftharpoons Cr^{3+}(aq) + 3OH^-(aq)$

$K_{sp} = [Cr^{3+}][OH^-]^3$

c. $Ba(OH)_2(s) \rightleftharpoons Ba^{2+}(aq) + 2OH^-(aq)$

$K_{sp} = [Ba^{2+}][OH^-]^2$

d. $Sn(OH)_2(s) \rightleftharpoons Sn^{2+}(aq) + 2OH^-(aq)$

$K_{sp} = [Sn^{2+}][OH^-]^2$

110. molar mass AgCl = 143.4 g

$9.0 \times 10^{-4} \, \text{g AgCl/L} \times \dfrac{1 \text{ mol AgCl}}{143.4 \text{ g AgCl}} = 6.28 \times 10^{-6} \text{ mol AgCl/L}$

$AgCl(s) \rightleftharpoons Ag^+(aq) + Cl^-(aq)$

$K_{sp} = [Ag^+][Cl^-] = (6.28 \times 10^{-6} \, M)(6.28 \times 10^{-6} \, M) = 3.9 \times 10^{-11}$

112. molar mass $Ni(OH)_2$ = 92.71 g

$\dfrac{0.14 \text{ g Ni(OH)}_2}{1 \text{ L}} \times \dfrac{1 \text{ mol}}{92.71 \text{ g Ni(OH)}_2} = 1.510 \times 10^{-3} \, M$

$Ni(OH)_2(s) \rightleftharpoons Ni^{2+}(aq) + 2OH^-(aq)$

$K_{sp} = [Ni^{2+}][OH^-]^2$

If $1.510 \times 10^{-3} \, M$ of $Ni(OH)_2$ dissolves, then $[Ni^{2+}] = 1.510 \times 10^{-3} \, M$ and $[OH^-] = 2 \times (1.510 \times 10^{-3} \, M) = 3.020 \times 10^{-3} \, M$.

$K_{sp} = (1.510 \times 10^{-3} \, M)(3.020 \times 10^{-3} \, M)^2 = 1.4 \times 10^{-8}$

Chapter 17:    Equilibrium

114. The activation energy is the minimum energy two colliding molecules must possess in order for the collision to result in reaction. If molecules do not possess energies equal to or greater than $E_a$, a collision between these molecules will not result in a reaction.

116. Once a system has reached equilibrium the net concentration of product no longer increases because molecules of product already present react to form the original reactants. This is not to say that the *same* product molecules are necessarily always present.

118. $K = \dfrac{[CO]^2[O_2]}{[CO_2]^2} = \dfrac{[0.11\ M]^2[0.055\ M]}{[1.4\ M]^2} = 3.4 \times 10^{-4}$

120. $3H_2(g) + N_2(g) \rightleftharpoons 2NH_3(g)$

$K = \dfrac{[NH_3(g)]^2}{[N_2(g)][H_2(g)]^3} = \dfrac{\left[\dfrac{1.16\ mol}{3.50\ L}\right]^2}{\left[\dfrac{1.14\ mol}{3.50\ L}\right]\left[\dfrac{2.40\ mol}{3.50\ L}\right]^3} = 1.05$

122. $3O_2(g) \rightleftharpoons 2O_3(g)$

$K = \dfrac{[O_3(g)]^2}{[O_2(g)]^3}$

$1.8 \times 10^{-7} = \dfrac{[x]^2}{[0.062\ M]^3}$

$x = [O_3(g)] = \sqrt{4.3 \times 10^{-11}\ M} = 6.5 \times 10^{-6}\ M$

124. $H_2(g) + F_2(g) \rightleftharpoons 2HF(g)$

$K = \dfrac{[HF(g)]^2}{[H_2(g)][F_2(g)]}$

$2.1 \times 10^{-3} = \dfrac{[x]^2}{[0.083\ M][0.083\ M]}$

$x = [HF(g)] = \sqrt{1.447 \times 10^{-5}\ M} = 3.8 \times 10^{-3}\ M$

126.

|  | $N_2$ | $H_2$ | $NH_3$ |
|---|---|---|---|
| **Add $N_2$** | Increase | Decrease | Increase |
| **Remove $H_2$** | Increase | Decrease | Decrease |
| **Add $NH_3$** | Increase | Increase | Increase |
| **Add Ne** | No change | No change | No change |
| **Increase T** | Increase | Increase | Decrease |
| **Decrease V** | Decrease | Decrease | Increase |
| **Add catalyst** | No change | No change | No change |

# CUMULATIVE REVIEW

# Chapters 16 and 17

2. A conjugate acid–base pair consists of two species related to each other by donation or acceptance of a single proton, $H^+$. An acid has one more $H^+$ than its conjugate base; a base has one less $H^+$ than its conjugate acid.

   Brønsted-Lowry acids:

   $HCl(aq) + H_2O(l) \rightarrow Cl^-(aq) + H_3O^+(aq)$

   $H_2SO_4(aq) + H_2O(l) \rightarrow HSO_4^-(aq) + H_3O^+(aq)$

   $H_3PO_4(aq) + H_2O(l) \rightleftharpoons H_2PO_4^-(aq) + H_3O^+(aq)$

   $NH_4^+(aq) + H_2O(l) \rightleftharpoons NH_3(aq) + H_3O^+(aq)$

   Brønsted-Lowry bases:

   $NH_3(aq) + H_2O(l) \rightleftharpoons NH_4^+(aq) + OH^-(aq)$

   $HCO_3^-(aq) + H_2O(l) \rightleftharpoons H_2CO_3(aq) + OH^-(aq)$

   $NH_2^-(aq) + H_2O(l) \rightarrow NH_3(aq) + OH^-(aq)$

   $H_2PO_4^-(aq) + H_2O(l) \rightleftharpoons H_3PO_4(aq) + OH^-(aq)$

4. The strength of an acid is a direct result of the position of the acid's ionization equilibrium. Strong acids are those whose ionization equilibrium positions lie far to the right, whereas weak acids are those whose equilibrium positions lie only slightly to the right. For example, HCl, $HNO_3$, and $HClO_4$ are all strong acids, which means they are completely ionized in aqueous solution (the position of equilibrium is very far to the right):

   $HCl(aq) + H_2O(l) \rightarrow Cl^-(aq) + H_3O^+(aq)$

   $HNO_3(aq) + H_2O(l) \rightarrow NO_3^-(aq) + H_3O^+(aq)$

   $HClO_4(aq) + H_2O(l) \rightarrow ClO_4^-(aq) + H_3O^+(aq)$

   As these are very strong acids, we know their anions ($Cl^-$, $NO_3^-$, $ClO_4^-$) must be very weak bases, and that solutions of the sodium salts of these anions would *not* be appreciably basic. As these acids have a strong tendency to lose protons, there is very little tendency for the anions (bases) to gain protons.

6. The pH of a solution is defined as the negative of the base 10 logarithm of the hydrogen ion concentration in the solution; that is

   $pH = -\log_{10}[H^+]$.

   As in pure water, the amount of $H^+(aq)$ ion present is equal to the amount of $OH^-(aq)$ ion, we say that pure water is *neutral*. As $[H^+] = 1.0 \times 10^{-7}$ $M$ in pure water, this means that the pH of pure water is $-\log[1.0 \times 10^{-7}\ M] = 7.00$. Solutions in which the hydrogen ion concentration is greater than $1.0 \times 10^{-7}\ M$ (pH < 7.00) are *acidic*; solutions in which the hydrogen ion concentration is less than $1.0 \times 10^{-7}$ M (pH > 7.00) are *basic*. The pH scale is logarithmic. When the pH changes by one unit, this corresponds to a change in the hydrogen ion concentration by a factor of *ten*.

In some instances, it may be more convenient to speak directly about the hydroxide ion concentration present in a solution, and so an analogous logarithmic expression is defined for the hydroxide ion concentration:

$$pOH = -\log_{10}[OH^-].$$

The concentrations of hydrogen ion and hydroxide ion in water (and in aqueous solutions) are *not* independent of one another, but rather are related by the dissociation equilibrium constant for water,

$$K_w = [H^+][OH^-] = 1.0 \times 10^{-14} \text{ at } 25°C.$$

From this constant it is obvious that pH + pOH = 14.00 for water (or an aqueous solution) at 25°C.

8. Chemists envision that a reaction can only take place between molecules if the molecules physically *collide* with each other. Furthermore, when molecules collide, the molecules must collide with enough force for the reaction to be successful (there must be enough energy to break bonds in the reactants), and the colliding molecules must be positioned with the correct relative orientation for the products (or intermediates) to form. Reactions tend to be faster if higher concentrations are used for the reaction; because, if there are more molecules present per unit volume there will be more collisions between molecules in a given time period. Reactions are faster at higher temperatures because at higher temperatures the reactant molecules have a higher average kinetic energy, and the number of molecules that will collide with sufficient force to break bonds increases.

10. Chemists define equilibrium as the exact balancing of two exactly opposing processes. When a chemical reaction is begun by combining pure reactants, the only process possible initially is

       reactants → products

However, for many reactions, as the concentration of product molecules increases, it becomes more likely that product molecules will collide and react with each other

       products → reactants

giving back molecules of the original reactants. At some point in the process the rates of the forward and reverse reactions become equal, and the system attains chemical equilibrium. To an outside observer, the system appears to have stopped reacting. On a microscopic basis, though, both the forward and reverse processes are still going on: every time additional molecules of the product form, however, somewhere else in the system molecules of product react to give back molecules of reactant.

Once the point is reached that product molecules are reacting at the same speed at which they are forming, there is no further net change in concentration. A graph showing how the rates of the forward and reverse reactions change with time is given in the text as Figure 17.8. At the start of the reaction, the rate of the forward reaction is at its maximum, whereas the rate of the reverse reaction is zero. As the reaction proceeds, the rate of the forward reaction gradually decreases as the concentration of reactants decreases, whereas the rate of the reverse reaction increases as the concentration of products increases. Once the two rates have become equal, the reaction has reached a state of equilibrium.

12. The equilibrium constant for a reaction is a *ratio* of the concentration of products present at the point of equilibrium to the concentration of reactants still present. A *ratio* means that we have one number divided by another number (for example, the density of a substance is the ratio of a substance's mass to its volume). As the equilibrium constant is a ratio, there are an infinite number of sets of data that can give the same ratio: for example, the ratios 8/4, 6/3, 100/50 all have the same value, 2. The actual concentrations of products and reactants will differ from one experiment to another involving a particular chemical reaction, but the ratio of the amount of product to reactant at equilibrium should be the same for each experiment.

Consider this simple example: suppose we have a reaction for which $K = 4$, and we begin this reaction with 100 reactant molecules. At the point of equilibrium, there should be 80 molecules of product and 20 molecules of reactant remaining (80/20 = 4). Suppose we perform another experiment involving the same reaction, only this time we begin the experiment with 500 molecules of reactant. This time, at the point of equilibrium, there will be 400 molecules of product present and 100 molecules of reactant remaining (400/100 = 4). As we began the two experiments with different numbers of reactant molecules, it's not troubling that there are different absolute numbers of product and reactant molecules present at equilibrium; however, the ratio, $K$, is the same for both experiments. We say that these two experiments represent two different positions of equilibrium: an equilibrium position corresponds to a particular set of experimental equilibrium concentrations that fulfill the value of the equilibrium constant. Any experiment that is performed with a different amount of starting material will come to its own unique equilibrium position, but the equilibrium constant ratio, $K$, will be the same for a given reaction regardless of the starting amounts taken.

14. Your paraphrase of Le Châtelier's principle should go something like this: "when you make any change to a system in equilibrium, this throws the system temporarily out of equilibrium, and the system responds by reacting in whichever direction will be able to reach a new position of equilibrium". There are various changes that can be made to a system in equilibrium. Following are examples:

   a. The concentration of one of the reactants is increased.

   Consider the reaction: $2SO_2(g) + O_2(g) \rightleftharpoons 2SO_3(g)$

   Suppose the reactants have already reacted and a position of equilibrium has been reached that fulfills the value of $K$ for the reaction. At this point there will be present particular amounts of each reactant and of product. Suppose then one additional mole of $O_2$ is added to the system from outside. At the instant the additional $O_2$ is added, the system will not be in equilibrium: there will be too much $O_2$ present in the system to be compatible with the amounts of $SO_2$ and $SO_3$ present. The system will respond by reacting to get rid of some of the excess $O_2$ until the value of the ratio $K$ is again fulfilled. If the system reacts to get rid of the excess of $O_2$, additional product $SO_3$ will form. The net result is more $SO_3$ produced than if the change had not been made.

   b. The concentration of one of the products is decreased by selectively removing it from the system.

   Consider the reaction: $CH_3COOH + CH_3OH \rightleftharpoons H_2O + CH_3COOCH_3$

   This reaction is typical of many reactions involving organic chemical substances, in which two organic molecules react to form a larger molecule, with a molecule of water split out during the combination. This type of reaction on its own tends to come to equilibrium with only part of the starting materials being converted to the desired organic product (which effectively would leave the experimenter with a mixture of materials). A technique that is used by organic chemists to increase the effective yield of the desired organic product is to *separate* the two products (if the products are separated, they cannot react to give back the reactants). One method used is to add a drying agent to the mixture: such a drying agent chemically or physically absorbs the water from the system, removing it from equilibrium. If the water is removed, the reverse reaction cannot take place, and the reaction proceeds to a greater extent in the forward direction than if the drying agent had not been added. In other situations, an experimenter may separate the products of the reaction by distillation (if the boiling points make this possible): again, if the products have been separated, then the reverse reaction will not be possible, and the forward reaction will occur to a greater extent.

c. The reaction system is compressed to a smaller volume.

Consider the example: $3H_2(g) + N_2(g) \rightleftharpoons 2NH_3(g)$

For equilibria involving gases, when the volume of the reaction system is compressed suddenly, the pressure in the system increases. However, if the reacting system can relieve some of this increased pressure by reacting, it will do so. This will happen by the reaction occurring in whichever direction will give the smaller number of moles of gas (if the number of moles of gas is decreased in a particular volume, the pressure will decrease).

For the reaction above, there are two moles of the gas on the right side of the equation, but there is a total of four moles on the left side. If this system at equilibrium were to be suddenly compressed to a smaller volume, the reaction would proceed further to the right (in favor of more ammonia being produced).

d. The temperature is increased for an endothermic reaction.

Consider the reaction: $2NaHCO_3 + heat \rightleftharpoons Na_2CO_3 + H_2O + CO_2$

Although a change in temperature actually does change the *value* of the equilibrium constant, we can simplify reactions involving temperature changes by treating heat energy as if it were a chemical substance: for this endothermic reaction, heat is one of the reactants. As we saw in the example in part (a) of this question, increasing the concentration of one of the reactants for a system at equilibrium causes the reaction to proceed further to the right, forming additional product. Similarly for the endothermic reaction given above, increasing the temperature causes the reaction to proceed further in the direction of products than if no change had been made. It is as if there were too much "heat" to be compatible with the amount of substances present. The substances react to get rid of some of the energy.

e. The temperature is decreased for an exothermic process.

Consider the reaction: $PCl_3 + Cl_2 \rightleftharpoons PCl_5 + heat$

As discussed in part (d) above, although changing the temperature at which a reaction is performed does change the numerical value of $K$, we can simplify our discussion of this reaction by treating heat energy as if it were a chemical substance. Heat is a product of this reaction. If we are going to lower the temperature of this reaction system, the only way to accomplish this is to remove energy from the system. Lowering the temperature of the system is really working with this system in its attempt to release heat energy. So lowering the temperature should favor the production of more product than if no change were made.

Review: Chapters 16 and 17

16. Specific answer depends on student choice of examples. In general, for a weak acid, HA, and a weak base, B:

$$HA + H_2O \rightleftharpoons H_3O^+ + A^- \qquad B + H_2O \rightleftharpoons HB^+ + OH^-$$

18.  a. $NH_3(aq)$(base) + $H_2O(l)$(acid) $\rightleftharpoons$ $NH_4^+(aq)$(acid) + $OH^-(aq)$(base)

   b. $H_2SO_4(aq)$(acid) + $H_2O(l)$(base) $\rightleftharpoons$ $HSO_4^-(aq)$(base) + $H_3O^+(aq)$(acid)

   c. $O^{2-}(s)$(base) + $H_2O(l)$(acid) $\rightleftharpoons$ $OH^-(aq)$(acid) + $OH^-(aq)$(base)

   d. $NH_2^-(aq)$(base) + $H_2O(l)$(acid) $\rightleftharpoons$ $NH_3(aq)$(acid) + $OH^-(aq)$(base)

   e. $H_2PO_4^-(aq)$(acid) + $OH^-(aq)$(base) $\rightleftharpoons$ $HPO_4^{2-}(aq)$(base) + $H_2O(l)$(acid)

20.  a. $HClO_4$ is a strong acid, so $[H^+] = 0.00562\ M$

   pH = $-\log(0.00562) = 2.250$

   pOH = $14.000 - 2.250 = 11.750$

   b. KOH is a strong base, so $[OH^-] = 3.98 \times 10^{-4}\ M$

   pOH = $-\log(3.98 \times 10^{-4}) = 3.400$

   pH = $14.000 - 3.400 = 10.600$

   c. $HNO_3$ is a strong acid, so $[H^+] = 0.078\ M$

   pH = $-\log(0.078) = 1.11$

   pOH = $14.00 - 1.11 = 12.89$

   d. $Ca(OH)_2$ is a strong, but not very soluble base. Each formula unit of $Ca(OH)_2$ produces two formula units of $OH^-$ ion.

   $[OH^-] = 2 \times 4.71 \times 10^{-6}\ M = 9.42 \times 10^{-6}\ M$

   pOH = $-\log(9.42 \times 10^{-6}) = 5.026$

   pH = $14.000 - 5.026 = 8.974$

22. $Br_2(g) + Cl_2(g) \rightleftharpoons 2BrCl(g)$

$$K = \frac{[BrCl(g)]^2}{[Br_2][Cl_2]} = \frac{[4.9 \times 10^{-4}]^2}{[7.2 \times 10^{-8}][4.3 \times 10^{-6}]} = 7.8 \times 10^5$$

24. $MgCO_3 \rightleftharpoons Mg^{2+}(aq) + CO_3^{2-}(aq)$ \qquad molar mass of $MgCO_3$ = 84.32 g

$K_{sp} = [Mg^{2+}][CO_2^{2-}]$

let $x$ represent the number of moles of $MgCO_3$ that dissolve per liter. Then $[Mg^{2+}] = x$ and $[CO_3^{2-}] = x$ also

$K_{sp} = [x][x] = x^2 = 6.82 \times 10^{-6}$

$x = [MgCO_3] = 2.61 \times 10^{-3}\ M$

$$\frac{2.61 \times 10^{-3}\ \text{mol}}{1\ L} \times \frac{84.32\ g}{1\ \text{mol}} = 0.220\ \text{g/L}$$

# CHAPTER 18

# Oxidation–Reduction Reactions and Electrochemistry

2. Oxidation can be defined as the loss of electrons by an atom, molecule, or ion. Oxidation may also be defined as an increase in oxidation state for an element, but because elements can only increase their oxidation states by losing electrons, the two definitions are equivalent. The following equation shows the oxidation of copper metal to copper(II) ion

   $$Cu \rightarrow Cu^{2+} + 2e^-$$

   Reduction can be defined as the gaining of electrons by an atom, molecule, or ion. Reduction may also be defined as a decrease in oxidation state for an element, but naturally such a decrease takes place by the gaining of electrons (so the two definitions are equivalent). The following equation shows the reduction of sulfur atoms to sulfide ion.

   $$S + 2e^- \rightarrow S^{2-}$$

4. Each of these reactions involves one or more *free* elements on one side of the equation; on the other side of the equation, however, the element(s) is(are) *combined* in a compound. This is a clear sign that an oxidation–reduction process is taking place.

   a. sodium is oxidized; nitrogen is reduced

   b. magnesium is oxidized; chlorine is reduced

   c. aluminum is oxidized; bromine is reduced

   d. magnesium is oxidized; copper is reduced

6. Each of these reactions involves one or more *free* elements on one side of the equation; on the other side of the equation, however, the element(s) is(are) *combined* in a compound. This is a clear sign that an oxidation–reduction process is taking place.

   a. magnesium is oxidized; bromine is reduced

   b. sodium is oxidized; sulfur is reduced

   c. hydrogen is oxidized; carbon is reduced

   d. potassium is oxidized; nitrogen is reduced

8. A neutral molecule has an overall charge of zero.

10. Fluorine is always assigned a negative oxidation state (–1) because all other elements are less electronegative. The other halogens are *usually* assigned an oxidation state of –1 in compounds. In interhalogen compounds such as ClF, fluorine is assigned oxidation state –1 (F is more electronegative than Cl). Chlorine, therefore, must be assigned a +1 oxidation state in this instance.

Chapter 18:   Oxidation–Reduction Reactions and Electrochemistry

12. The sum of all the oxidation states of the atoms in a polyatomic ion must equal the overall charge on the ion. The sum of all the oxidation states of all the atoms in $PO_4^{3-}$ is $-3$.

14. The rules for assigning oxidation states are given in Section 18.2 of the text. The rule that applies for each element in the following answers is given in parentheses after the element and its oxidation state.

   a.   N, +3 (Rule 6); Br, –1 (Rule 5)

   b.   Se, +6 (Rule 6); F, –1 (Rule 5)

   c.   P, +5 (Rule 6); Br, –1 (Rule 5)

   d.   C, –4 (Rule 6); H, +1 (Rule 4)

16. a.   0 (Rule 1)

   b.   –3 (using Rule 4 for H)

   c.   +4 (using Rule 3 for O)

   d.   +5 (using Rule 3 for O and Rule 2 for Na)

18. a.   +2 (using Rule 5 for Cl)

   b.   +7 (using Rule 3 for O and Rule 2 for K)

   c.   +4 (using Rule 3 for O)

   d.   +3 (realizing that the acetate ion has 1– charge and apply Rule 6)

20. The rules for assigning oxidation states are given in Section 18.2 of the text. The rule that applies for each element in the following answers is given in parentheses after the element and its oxidation state.

   a.   Mg, +2 (Rule 6); O, –2 (Rule 3)

   b.   Fe, +3 (Rules 6 and 2); O, –2 (Rule 3)

   c.   P, +3 (Rule 6); Cl, –1 (Rule 5)

   d.   N, +5 (Rule 6); O, –2 (Rule 3)

22. The rules for assigning oxidation states are given in Section 18.2 of the text. The rule that applies for each element in the following answers is given in parentheses after the element and its oxidation state.

   a.   H, +1 (Rule 4); N, –3 (Rule 7)

   b.   H, +1 (Rule 4); C, +4 (Rule 7); O, –2 (Rule 3)

   c.   H, +1 (Rule 4); O, –2 (Rules 3 and 7)

   d.   Cr, +6 (Rule 7); O, –2 (Rule 3)

24. Electrons are negative; when an atom gains electrons, it gains one negative charge for each electron gained. For example, in the reduction reaction $Cl + e^- \rightarrow Cl^-$, the oxidation state of chlorine decreases from 0 to –1 as the electron is gained.

26. An oxidizing agent decreases its oxidation state. A reducing agent increases its oxidation state.

Chapter 18: Oxidation–Reduction Reactions and Electrochemistry

28. An antioxidant is a substance that prevents oxidation of some molecule(s) in the body. It is not certain how all antioxidants work, but one example is in preventing oxygen molecules and other substances from stripping electrons from cell membranes, leaving them vulnerable to destruction by the immune system.

30. a. $2Al(s) + 3S(s) \rightarrow Al_2S_3(s)$
aluminum is being oxidized, sulfur is being reduced

b. $CH_4(g) + 2O_2(g) \rightarrow CO_2(g) + 2H_2O(g)$
carbon is being oxidized, oxygen is being reduced

c. $2Fe_2O_3(s) + 3C(s) \rightarrow 3CO_2(g) + 4Fe(s, l)$
carbon is being oxidized, iron is being reduced

d. $K_2Cr_2O_7(aq) + 14HCl(aq) \rightarrow 2KCl(aq) + 2CrCl_3(s) + 7H_2O(l) + 3Cl_2(g)$
chlorine is being oxidized, chromium is being reduced

32. a. $4KClO_3(s) + C_6H_{12}O_6(s) \rightarrow 4KCl(s) + 6H_2O(l) + 6CO_2(g)$
carbon is being oxidized, chlorine is being reduced

b. $2C_8H_{18}(l) + 25O_2(g) \rightarrow 16CO_2(g) + 18H_2O(l)$
carbon is being oxidized, oxygen is being reduced

c. $PCl_3(g) + Cl_2(g) \rightarrow PCl_5(g)$
phosphorus is being oxidized, chlorine is being reduced

d. $Ca(s) + H_2(g) \rightarrow CaH_2(g)$
calcium is being oxidized, hydrogen is being reduced

34. Iron is reduced [+3 in $Fe_2O_3(s)$, 0 in $Fe(l)$]; carbon is oxidized [+2 in $CO(g)$, +4 in $CO_2(g)$]. $Fe_2O_3(s)$ is the oxidizing agent; $CO(g)$ is the reducing agent.

36. a. chlorine is being reduced, iodine is being oxidized; chlorine is the oxidizing agent, iodide ion is the reducing agent

b. iron is being reduced, iodine is being oxidized; iron(III) is the oxidizing agent, iodide ion is the reducing agent

c. copper is being reduced, iodine is being oxidized; copper(II) is the oxidizing agent, iodide ion is the reducing agent

38. Oxidation–reduction reactions are often more complicated than "regular" reactions; frequently the coefficients necessary to balance the number of electrons transferred come out to be large numbers. We also have to make certain that we account for the electrons being transferred.

40. Under ordinary conditions it is impossible to have "free" electrons that are not part of some atom, ion, or molecule. For this reason, the total number of electrons lost by the species being oxidized must equal the total number of electrons gained by the species being reduced.

42. a. $N_2(g) \rightarrow N_3^-(aq)$

balance nitrogen: **$3N_2(g) \rightarrow 2N_3^-(aq)$**

## Chapter 18: Oxidation–Reduction Reactions and Electrochemistry

        balance charge: $3N_2(g) + 2e^- \rightarrow 2N_3^-(aq)$

        balanced half-reaction: $3N_2(g) + 2e^- \rightarrow 2N_3^-(aq)$

    b.   $O_2^{2-}(aq) \rightarrow O_2(g)$

        balance charge: $O_2^{2-}(aq) \rightarrow O_2(g) + 2e^-$

        balanced half-reaction: $O_2^{2-}(aq) \rightarrow O_2(g) + 2e^-$

    c.   $Zn(s) \rightarrow Zn^{2+}(aq)$

        balance charge: $Zn(s) \rightarrow Zn^{2+}(aq) + 2e^-$

        balanced half-reaction: $Zn(s) \rightarrow Zn^{2+}(aq) + 2e^-$

    d.   $F_2(g) \rightarrow F^-(aq)$

        balance flourine: $F_2(g) \rightarrow 2F^-(aq)$

        balance charge: $F_2(g) + 2e^- \rightarrow 2F^-(aq)$

        balanced half-reaction: $F_2(g) + 2e^- \rightarrow 2F^-(aq)$

44.   a.   $O_2(g) \rightarrow H_2O(l)$

        balance oxygen: $O_2 \rightarrow 2H_2O$

        balance hydrogen: $4H^+ + O_2 \rightarrow 2H_2O$

        balance charge: $4e^- + 4H^+ + O_2 \rightarrow 2H_2O$

        balanced half-reaction: $4e^- + 4H^+(aq) + O_2(g) \rightarrow 2H_2O(l)$

    b.   $SO_4^{2-}(aq) \rightarrow H_2SO_3(aq)$

        balance oxygen: $SO_4^{2-} \rightarrow H_2SO_3 + H_2O$

        balance hydrogen: $4H^+ + SO_4^{2-} \rightarrow H_2SO_3 + H_2O$

        balance charge: $2e^- + 4H^+ + SO_4^{2-} \rightarrow H_2SO_3 + H_2O$

        balanced half-reaction: $2e^- + 4H^+(aq) + SO_4^{2-}(aq) \rightarrow H_2SO_3(aq) + H_2O(l)$

    c.   $H_2O_2(aq) \rightarrow H_2O(l)$

        balance oxygen : $H_2O_2 \rightarrow 2H_2O$

        balance hydrogen : $2H^+ + H_2O_2 \rightarrow 2H_2O$

        balance charge : $2e^- + 2H^+ + H_2O_2 \rightarrow 2H_2O$

        balanced half-reaction: $2e^- + 2H^+(aq) + H_2O_2(aq) \rightarrow 2H_2O(l)$

    d.   $NO_2^-(aq) \rightarrow NO_3^-(aq)$

        balance oxygen : $H_2O + NO_2^- \rightarrow NO_3^-$

        balance hydrogen: $H_2O + NO_2^- \rightarrow NO_3^- + 2H^+$

        balance charge: $H_2O + NO_2^- \rightarrow NO_3^- + 2H^+ + 2e^-$

        balanced half-reaction: $H_2O(l) + NO_2^-(aq) \rightarrow NO_3^-(aq) + 2H^+(aq) + 2e^-$

Chapter 18: Oxidation–Reduction Reactions and Electrochemistry

46. For simplicity, the physical states of the substances have been omitted until the final balanced equation is given.

   a. $Al(s) + H^+(aq) \rightarrow Al^{3+}(aq) + H_2(g)$

   $Al \rightarrow Al^{3+}$

   Balance charge: $Al \rightarrow Al^{3+} + \mathbf{3e^-}$

   $H^+ \rightarrow H_2$

   Balance hydrogen: $\mathbf{2H^+} \rightarrow H_2$

   Balance charge: $\mathbf{2e^-} + 2H^+ \rightarrow H_2$

   Combine half–reactions:

   $3 \times (2e^- + 2H^+ \rightarrow H_2)$

   $2 \times (Al \rightarrow Al^{3+} + 3e^-)$

   $2Al(s) + 6H^+(aq) \rightarrow 2Al^{3+}(aq) + 3H_2(g)$

   b. $S^{2-}(aq) + NO_3^-(g) \rightarrow S(s) + NO(g)$

   $S^{2-} \rightarrow S$

   Balance charge: $S^{2-} \rightarrow S + \mathbf{2e^-}$

   $NO_3^- \rightarrow NO$

   Balance oxygen: $NO_3^- \rightarrow NO + 2H_2O$

   Balance hydrogen: $\mathbf{4H^+} + NO_3^- \rightarrow NO + 2H_2O$

   Balance charge: $\mathbf{3e^-} + 4H^+ + NO_3^- \rightarrow NO + 2H_2O$

   Combine half–reactions:

   $3 \times (S^{2-} \rightarrow S + 2e^-)$

   $2 \times (3e^- + 4H^+ + NO_3^- \rightarrow NO + 2H_2O)$

   $8H^+ + 3S^{2-}(aq) + 2NO_3^-(g) \rightarrow 3S(s) + 2NO(g) + 4H_2O$

   c. $I_2(aq) + Cl_2(aq) \rightarrow IO_3^-(aq) + HCl(g)$

   $I_2 \rightarrow IO_3^-$

   Balance iodine: $I_2 \rightarrow \mathbf{2IO_3^-}$

   Balance oxygen: $\mathbf{6H_2O} + I_2 \rightarrow 2IO_3^-$

   Balance hydrogen: $6H_2O + I_2 \rightarrow 2IO_3^- + \mathbf{12H^+}$

   Balance charge: $6H_2O + I_2 \rightarrow 2IO_3^- + 12H^+ + \mathbf{10e^-}$

   $Cl_2 \rightarrow HCl$

   Balance chlorine: $Cl_2 \rightarrow \mathbf{2HCl}$

   Balance hydrogen: $\mathbf{2H^+} + Cl_2 \rightarrow 2HCl$

   Balance charge: $\mathbf{2e^-} + 2H^+ + Cl_2 \rightarrow 2HCl$

   Combine half–reactions:

# Chapter 18: Oxidation–Reduction Reactions and Electrochemistry

$$5 \times (2e^- + 2H^+ + Cl_2 \rightarrow 2HCl)$$

$$6H_2O + I_2 \rightarrow 2IO_3^- + 12H^+ + 10e^-$$

$$6H_2O(l) + 2I_2(aq) + 5Cl_2(aq) \rightarrow 2IO_3^-(aq) + 10HCl(g) + 2H^+(aq)$$

d. $AsO_4^-(aq) + S^{2-}(aq) \rightarrow AsO_3^-(s) + S(s)$

$AsO_4^- \rightarrow AsO_3^-$

Balance oxygen: $AsO_4^- \rightarrow AsO_3^- + \mathbf{H_2O}$

Balance hydrogen: $\mathbf{2H^+} + AsO_4^- \rightarrow AsO_3^- + H_2O$

Balance charge: $\mathbf{2e^-} + 2H^+ + AsO_4^- \rightarrow AsO_3^- + H_2O$

$S^{2-} \rightarrow S$

Balance charge: $S^{2-} \rightarrow S + \mathbf{2e^-}$

$2H^+(aq) + AsO_4^-(aq) + S^{2-}(aq) \rightarrow AsO_3^-(s) + S(s) + H_2O(l)$

48. $Cu(s) + 2HNO_3(aq) + 2H^+(aq) \rightarrow Cu^{2+}(aq) + 2NO_2(g) + 2H_2O(l)$

$Mg(s) + 2HNO_3(aq) \rightarrow Mg(NO_3)_2(aq) + H_2(g)$

50. A salt bridge typically consists of a *U*–shaped tube filled with an inert electrolyte (one involving ions that are not part of the oxidation–reduction reaction). A salt bridge is used to complete the electrical circuit in a cell. Any method that allows transfer of charge without allowing bulk mixing of the solutions may be used (another common method is to set up one half–cell in a porous cup, which is then placed in the beaker containing the second half–cell).

52. Reduction takes place at the cathode and oxidation takes place at the anode.

54. A diagram of the cell is shown below:

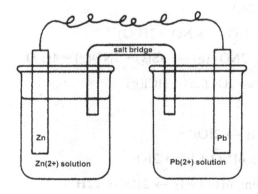

$Pb^{2+}(aq)$ ion is reduced; $Zn(s)$ is oxidized.

The reaction at the anode is $Zn(s) \rightarrow Zn^{2+}(aq) + 2e^-$.

The reaction at the cathode is $Pb^{2+}(aq) + 2e^- \rightarrow Pb(s)$

56. Some advantages include: the many different shapes available to accommodate various types of electronic devices; they are lighter than competing types of batteries; and they have no memory effect. One disadvantage is that if the battery is overheated or overcharged, it can rupture and cause a fire.

Chapter 18: Oxidation–Reduction Reactions and Electrochemistry

58. Aluminum is a very reactive metal when freshly isolated in the pure state. However, on standing for even a relatively short period of time, aluminum metal forms a thin coating of Al$_2$O$_3$ on its surface from reaction with atmospheric oxygen. This coating of Al$_2$O$_3$ is much less reactive than the metal and serves to protect the surface of the metal from further attack.

60. Chromium protects stainless steel by forming a thin coating of chromium oxide on the surface of the steel, which prevents oxidation of the iron in the steel.

62. Some important uses of electrolysis include rechargeable batteries (lead storage battery) and in the production of metals from their ores (such as aluminum).

64. The balanced equation is 2H$_2$O(l) → 2H$_2$(g) + O$_2$(g). Oxygen is oxidized (going from –2 oxidation state in water to zero oxidation state in the free element). Hydrogen is reduced (going from +1 oxidation state in water to zero oxidation state in the free element). Heat is produced by burning the hydrogen gas produced by the electrolysis: since energy must be applied to water to electrolyze it, energy is released when hydrogen gas produced by the electrolysis and oxygen gas combine to form water in the fireplace.

66. oxidation numbers

68. electronegative

70. An *oxidizing agent* is an atom, molecule, or ion that causes the oxidation of another species. During this process, the oxidizing agent is reduced.

72. lose

74. separate from

76. oxidation

78. galvanic

80. electrolysis; In a galvanic cell, chemical energy is converted to electrical energy by means of an oxidation-reduction reaction. In electrolysis, electrical energy is used to produce a chemical change.

82. oxidation

84. a.  4Fe(s) + 3O$_2$(g) → 2Fe$_2$O$_3$(s)

    iron is oxidized (and is therefore the reducing agent); oxygen is reduced

　　b.  2Al(s) + 3Cl$_2$(g) → 2AlCl$_3$(s)

    aluminum is oxidized (and is therefore the reducing agent); chlorine is reduced

　　c.  6Mg(s) + P$_4$(s) → 2Mg$_3$P$_2$(s)

    magnesium is oxidized (and is therefore the reducing agent); phosphorus is reduced

86. a.  Hydrogen is reduced (+1 → 0) therefore HCl is the oxidizing agent.

　　b.  Hydrogen is reduced (+1 → 0) therefore H$_2$SO$_4$ is the oxidizing agent.

## Chapter 18: Oxidation–Reduction Reactions and Electrochemistry

88.
    a.     $C_3H_8(g) + 5O_2(g) \rightarrow 3CO_2(g) + 4H_2O(g)$

    b.     $CO(g) + 2H_2(g) \rightarrow CH_3OH(l)$

    c.     $SnO_2(s) + 2C(s) \rightarrow Sn(s) + 2CO(g)$

    d.     $C_2H_5OH(l) + 3O_2(g) \rightarrow 2CO_2(g) + 3H_2O(g)$

90.     Each of these reactions involves a *metallic* element in the form of the *free* element on one side of the equation; on the other side of the equation, the metallic element is *combined* in an ionic compound. If a metallic element goes from the free metal to the ionic form, the metal is oxidized (loses electrons).

    a.     sodium is oxidized; oxygen is reduced

    b.     iron is oxidized; hydrogen is reduced

    c.     oxygen ($O^{2-}$) is oxidized; aluminum ($Al^{3+}$) is reduced (this reaction is the reverse of the type discussed above)

    d.     magnesium is oxidized; nitrogen is reduced

92.     The rules for assigning oxidation states are given in Section 18.2 of the text. The rule that applies for each element in the following answers is given in parentheses after the element and its oxidation state.

    a.     H +1 (Rule 4); N –3 (Rule 6)

    b.     C +2 (Rule 6); O –2 (Rule 3)

    c.     C +4 (Rule 6); O –2 (Rule 3)

    d.     N +3 (Rule 6); F –1 (Rule 5)

94.     The rules for assigning oxidation states are given in Section 18.2 of the text. The rule that applies for each element is that given in parentheses after the element and its oxidation state.

    a.     Mn +4 (Rule 6); O –2 (Rule 3)

    b.     Ba +2 (Rule 2); Cr +6 (Rule 6); O –2 (Rule 3)

    c.     H +1 (Rule 4); S +4 (Rule 6); O –2 (Rule 3)

    d.     Ca +2 (Rule 2); P +5 (Rule 6); O –2 (Rule 3)

96.     The rules for assigning oxidation states are given in Section 18.2 of the text. The rule that applies for each element is that given in parentheses after the element and its oxidation state.

    a.     Bi +3 (Rule 7); O –2 (Rule 3)

    b.     P +5 (Rule 7); O –2 (Rule 3)

    c.     N +3 (Rule 7); O –2 (Rule 3)

    d.     Hg +1 (Rule 7)

98.
    a.     $2B_2O_3(s) + 6Cl_2(g) \rightarrow 4BCl_3(l) + 3O_2(g)$

            oxygen is oxidized (–2 to 0); chlorine is reduced (0 to –1)

    b.     $GeH_4(g) + O_2(g) \rightarrow Ge(s) + 2H_2O(g)$

Chapter 18: Oxidation–Reduction Reactions and Electrochemistry

       germanium is oxidized (–4 to 0); oxygen is reduced (0 to –2)

   c.   $C_2H_4(g) + Cl_2(g) \rightarrow C_2H_4Cl_2(l)$

       carbon is oxidized –2 to –1); chlorine is reduced (0 to –1)

   d.   $O_2(g) + 2F_2(g) \rightarrow 2OF_2(g)$

       oxygen is oxidized (0 to +2); fluorine is reduced (0 to –1)

100.   a.   $SiO_2(s) \rightarrow Si(s)$

       Balance oxygen: $SiO_2(s) \rightarrow Si(s) + \mathbf{2H_2O}(l)$

       Balance hydrogen: $SiO_2(s) + \mathbf{4H^+}(aq) \rightarrow Si(s) + 2H_2O(l)$

       Balance charge: $SiO_2(s) + 4H^+(aq) + \mathbf{4e^-} \rightarrow Si(s) + 2H_2O(l)$

       Balanced half–reaction: $SiO_2(s) + 4H^+(aq) + 4e^- \rightarrow Si(s) + 2H_2O(l)$

   b.   $S(s) \rightarrow H_2S(g)$

       Balance hydrogen: $S(s) + \mathbf{2H^+}(aq) \rightarrow H_2S(g)$

       Balance charge: $S(s) + 2H^+(aq) + \mathbf{2e^-} \rightarrow H_2S(g)$

       Balanced half–reaction: $S(s) + 2H^+(aq) + 2e^- \rightarrow H_2S(g)$

   c.   $NO_3^-(aq) \rightarrow HNO_2(aq)$

       Balance oxygen: $NO_3^-(aq) \rightarrow HNO_2(aq) + \mathbf{H_2O}(l)$

       Balance hydrogen: $NO_3^-(aq) + \mathbf{3H^+}(aq) \rightarrow HNO_2(aq) + H_2O(l)$

       Balance charge: $NO_3^-(aq) + 3H^+(aq) + \mathbf{2e^-} \rightarrow HNO_2(aq) + H_2O(l)$

       Balanced half–reaction: $NO_3^-(aq) + 3H^+(aq) + 2e^- \rightarrow HNO_2(aq) + H_2O(l)$

   d.   $NO_3^-(aq) \rightarrow NO(g)$

       Balance oxygen: $NO_3^-(aq) \rightarrow NO(g) + \mathbf{2H_2O}(l)$

       Balance hydrogen: $NO_3^-(aq) + \mathbf{4H^+}(aq) \rightarrow NO(g) + 2H_2O(l)$

       Balance charge: $NO_3^-(aq) + 4H^+(aq) + \mathbf{3e^-} \rightarrow NO(g) + 2H_2O(l)$

       Balanced half–reaction: $NO_3^-(aq) + 4H^+(aq) + 3e^- \rightarrow NO(g) + 2H_2O(l)$

102.   The correct answer is *d*. $Al^{3+}(aq)$ ion is reduced. $Mg(s)$ is oxidized. Reduction occurs at the cathode and oxidation occurs at the anode. The reaction at the cathode is $2Al^{3+}(aq) + 6e^- \rightarrow 2Al(s)$. The reaction at the anode is $3Mg(s) \rightarrow 3Mg^{2+}(aq) + 6e^-$.

104.   Notice that both dyes include "$C_{16}N_2H_{10}O_2$". Since leucoindigo is $Na_2C_{16}N_2H_{10}O_2$, the "$C_{16}N_2H_{10}O_2$" portion has a 2– charge while indigo ($C_{16}N_2H_{10}O_2$) is neutral. Since the sum of the oxidation states equals the charge, the oxidation state of one or more of the elements must increase, thus the molecule must be oxidized.

CHAPTER 19

# Radioactivity and Nuclear Energy

2. The radius of a typical atomic nucleus is on the order of $10^{-13}$ cm, which is about one hundred thousand times smaller than the radius of an atom overall.

4. mass number

6. The general symbol for a nuclide is $^A_Z X$. The atomic number (Z) is written in such formulas as a left subscript, whereas the mass number (A) is written as a left superscript.

   Thus, for the nuclide $^{14}_6 C$, the mass number is 14 and the atomic number is 6.

8. When a nucleus produces an alpha particle, the atomic number of the parent nucleus decreases by two units.

10. Emission of a neutron, $^1_0 n$, does not change the atomic number of the parent nucleus, but causes the mass number of the parent nucleus to decrease by one unit.

12. Gamma rays are high-energy photons of electromagnetic radiation. They are not normally considered to be particles. When a nucleus produces only gamma radiation, the atomic number and mass number remain the same.

14. Electron capture occurs when one of the inner orbital electrons is pulled into and becomes part of the nucleus.

16. The fact that the average atomic mass of potassium is only slightly above 39 amu reflects the fact that the isotope of mass number 39 predominates.

    | Isotope | Number of neutrons |
    |---|---|
    | $^{39}_{19}K$ | 20 neutrons |
    | $^{40}_{19}K$ | 21 neutrons |
    | $^{41}_{19}K$ | 22 neutrons |

18. The approximate atomic molar mass could be calculated as follows:

    $0.79(24) + 0.10(25) + 0.11(26) = 24.3$.

    This is *only* an approximation because the mass numbers, rather than the actual isotopic masses, were used. The fact that the approximate mass calculated is slightly above 24 shows that the isotope of mass number 24 predominates.

20. The correct answer is *a*. In beta-particle production, $^{\ \ 0}_{-1} e$ is produced (e.g. $^{234}_{\ 90}Th \rightarrow ^{\ \ 0}_{-1}e + ^{234}_{\ 91}Pa$). The atomic number of the parent nuclide goes up, thus decreasing the neutron to proton ratio.

22. a. $^{196}_{\ 85}At$

Chapter 19: Radioactivity and Nuclear Energy

    b.    $^{208}_{84}Po$

    c.    $^{210}_{86}Rn$

24.    a.    $^{201}_{79}Au$

    b.    $^{210}_{82}Pb$

    c.    $^{210}_{84}Po$

26.    a.    $^{234}_{92}U \rightarrow {}^{4}_{2}He + {}^{230}_{90}Th$

    b.    $^{222}_{86}Rn \rightarrow {}^{4}_{2}He + {}^{218}_{84}Po$

    c.    $^{162}_{75}Re \rightarrow {}^{4}_{2}He + {}^{158}_{73}Ta$

28.    a.    $^{136}_{53}I \rightarrow {}^{0}_{-1}e + {}^{136}_{54}Xe$

    b.    $^{133}_{51}Sb \rightarrow {}^{0}_{-1}e + {}^{133}_{52}Te$

    c.    $^{117}_{49}In \rightarrow {}^{0}_{-1}e + {}^{117}_{50}Sn$

30.    In a nuclear bombardment process, a target nucleus is bombarded with high-energy particles (typically subatomic particles or small atoms) from a particle accelerator. This may result in the transmutation of the target nucleus into some other element. For example, nitrogen-14 may be transmuted into oxygen-17 by bombardment with alpha particles. There is often considerable repulsion between the target nucleus and the particles being used for bombardment (especially if the bombarding particle is positively charged like the target nucleus). Using accelerators to increases the kinetic energy of the bombarding particles can overcome this repulsion.

32.    $^{24}_{12}Mg + {}^{2}_{1}H \rightarrow {}^{22}_{11}Na + {}^{4}_{2}He$

34.    The half-life of a nucleus is the time required for one-half of the original sample of nuclei to decay. A given isotope of an element always has the same half-life, although different isotopes of the same element may have greatly different half-lives. Nuclei of different elements typically have different half-lives.

36.    $^{226}_{88}Ra$ is the most stable (longest half-life); $^{224}_{88}Ra$ is the "hottest" (shortest half-life)

38.    With a half-life of 2.8 hours, strontium-87 is the hottest; with a half-life of 45.1 days, iron-59 is the most stable to decay.

Chapter 19: Radioactivity and Nuclear Energy

40. Half-life, 1.5 min; let $x$ represent the starting amount of isotope

| time, min | 0 | 1.5 | 3.0 | 4.5 | 6.0 |
|---|---|---|---|---|---|
| mass | $x$ | $\frac{1}{2}x$ | $\frac{1}{4}x$ | $\frac{1}{8}x$ | $\frac{1}{16}x$ |

After six minutes (four half-lives), $\frac{1}{16}$ of the original Co-62 sample $[(\frac{1}{2})^4]$ will remain.

42. For an administered dose of 100 μg, 0.39 μg remains after 2 days. The fraction remaining is 0.39/100 = 0.0039; on a percentage basis, less than 0.4% of the original radioisotope remains.

44. Carbon-14 is produced in the upper atmosphere by the bombardment of ordinary nitrogen with neutrons from space:

$$^{14}_{7}N + ^{1}_{0}n \rightarrow ^{14}_{6}C + ^{1}_{1}H$$

46. We assume that the concentration of C-14 in the atmosphere is effectively constant. A living organism is constantly replenishing C-14 either through the processes of metabolism (sugars ingested in foods contain C-14), or photosynthesis (carbon dioxide contains C-14). When a plant dies, it can no longer replenish, and as the C-14 undergoes radioactive decay, its amount decreases with time.

48. 1 day is about 13 half-lives for $^{18}_{9}F$. If we begin with $6.02 \times 10^{23}$ atoms (1 mol), then after 13 half-lives, $7.4 \times 10^{19}$ atoms of $^{18}_{9}F$ will remain.

50. fission, fusion, fusion, fission

52. $^{1}_{0}n + ^{235}_{92}U \rightarrow ^{142}_{56}Ba + ^{91}_{36}Kr + 3^{1}_{0}n$ is one possibility.

54. A critical mass of a fissionable material is the amount needed to provide a high enough internal neutron flux to sustain the chain reaction (enough neutrons are produced to cause the continuous fission of further material). A sample with less than a critical mass is still radioactive, but cannot sustain a chain reaction.

56. An actual nuclear explosion, of the type produced by a nuclear weapon, cannot occur in a nuclear reactor because the concentration of the fissionable materials is not sufficient to form a supercritical mass. However, since many reactors are cooled by water, which can decompose into hydrogen and oxygen gases, a *chemical* explosion is possible that could scatter the radioactive material used in the reactor.

58. $^{238}_{92}U$; $^{239}_{94}Pu$

60. In one type of fusion reactor, two $^{2}_{1}H$ atoms are fused to produce $^{4}_{2}He$. Because the hydrogen nuclei are positively charged, extremely high energies (temperatures of 40 million K) are needed to overcome the repulsion between the nuclei as they are shot into each other.

62. protons (hydrogen); helium

64. Somatic damage is directly to the organism itself, causing nearly immediate sickness or death to the organism. Genetic damage is to the genetic machinery of the organism, which will be manifested in future generations of offspring.

Chapter 19: Radioactivity and Nuclear Energy

66. Gamma rays penetrate long distances, but seldom cause ionization of biological molecules. Alpha particles, because they are much heavier although less penetrating, are very effective at ionizing biological molecules and leave a dense trail of damage in the organism. Isotopes that release alpha particles can be ingested or breathed into the body where the damage from the alpha particles will be more acute.

68. Nuclear waste may remain radioactive for thousands of years, and much of it is chemically poisonous as well as radioactive. Most reactor waste is still in "temporary storage." Various suggestions have been made for a more permanent solution, such as casting the spent fuel into glass bricks to contain it, and then storing the bricks in corrosion-proof metal containers deep underground. No agreement on a permanent solution to the disposal of nuclear waste has yet been reached.

70. radioactive

72. atomic

74. neutron; proton

76. radioactive decay

78. alpha decay

80. transuranium

82. half-life

84. radiotracers

86. chain

88. Every 45 years, the sample has a mass half of what it had at the beginning of those 45 years. Over 225 years, the sample will be cut in half five times, so there must have been 128 g initially to have 4.00 g after 225 years.

90. a. $^{234}_{90}\text{Th}$; alpha-particle production

  b. $^{0}_{-1}\text{e}$; beta-particle production

92. $3.5 \times 10^{-11}$ J/atom; $8.9 \times 10^{10}$ J/g

94. $^{90}_{40}\text{Zr}$, $^{91}_{40}\text{Zr}$, $^{92}_{40}\text{Zr}$, $^{94}_{40}\text{Zr}$, and $^{96}_{40}\text{Zr}$

## Chapter 19: Radioactivity and Nuclear Energy

96. $^{27}_{13}\text{Al}$ (13 protons, 14 neutrons)

   $^{28}_{13}\text{Al}$ (13 protons, 15 neutrons)

   $^{29}_{13}\text{Al}$ (13 protons, 16 neutrons)

98. *Three* of the statements are true. Statements *a*, *b*, and *d* are true.

100. $^{9}_{4}\text{Be} + ^{4}_{2}\text{He} \rightarrow ^{12}_{6}\text{C} + ^{1}_{0}\text{n}$

102. $^{238}_{92}\text{U} + ^{1}_{0}\text{n} \rightarrow ^{239}_{92}\text{U}$

   $^{239}_{92}\text{U} \rightarrow ^{239}_{93}\text{Np} + ^{0}_{-1}\text{e}$

   $^{239}_{93}\text{Np} \rightarrow ^{239}_{94}\text{Pu} + ^{0}_{-1}\text{e}$

104. $^{4}_{2}\text{He}$; $^{0}_{-1}\text{e}$; $^{0}_{-1}\text{e}$; $^{0}_{-1}\text{e}$; $^{0}_{-1}\text{e}$

106. Half-life, 80.9 years; let *x* represent the starting amount of isotope

   | time, years | 0 | 80.9 | 162 (2×80.9) | 243 (3×80.9) |
   |---|---|---|---|---|
   | mass decayed | *x* | (50%)*x* | (75%)*x* | (87.5%)*x* |

   After 243 years (three half-lives), 87.5% of the substance has decayed.

# CHAPTER 20

# Organic Chemistry

2. Carbon has only four valence electrons and can only make 4 bonds to other atoms.

4. A triple bond represents the sharing of three pairs of electrons between two bonded atoms. The sharing of three pairs imparts a linear geometry in the region of the triple bond. The simplest example of an organic molecule containing a triple bond is acetylene, H–C≡C–H.

6. $\ddot{O}=C=\ddot{O}$    $C≡O$

8. Molecules a and c contain only carbon–carbon single bonds and are therefore saturated.

10. 109.5°

12. The general formula for the alkanes is $C_nH_{2n+2}$.
    a. 2(4) + 2 = 10
    b. 2(6) + 2 = 14
    c. 2(17) + 2 = 36
    d. 2(20) + 2 = 42

14. 
    a. pentane     $CH_3–CH_2–CH_2–CH_2–CH_3$
    b. undecane    $CH_3–CH_2–CH_2–CH_2–CH_2–CH_2–CH_2–CH_2–CH_2–CH_2–CH_3$
    c. nonane      $CH_3–CH_2–CH_2–CH_2–CH_2–CH_2–CH_2–CH_2–CH_3$
    d. heptane     $CH_3–CH_2–CH_2–CH_2–CH_2–CH_2–CH_3$

16. A branched alkane contains one or more shorter carbon atom chains, attached to the side of the main (longest) carbon atom chain. The simplest branched alkane is 2-methylpropane.

```
    H   H   H
    |   |   |
H—C—C—C—H
    |   |   |
    H  CH3  H
```

18. Carbon skeletons are shown.

## Chapter 20: Organic Chemistry

20. The root name is derived from the number of carbon atoms in the *longest continuous chain* of carbon atoms.

22. parent

24. prefix

26. Look for the *longest* continuous chain of carbon atoms.
    a. 3-ethylpentane
    b. 2,2-dimethylbutane
    c. 2,2-dimethylpropane
    d. 2,3,4-trimethylpentane

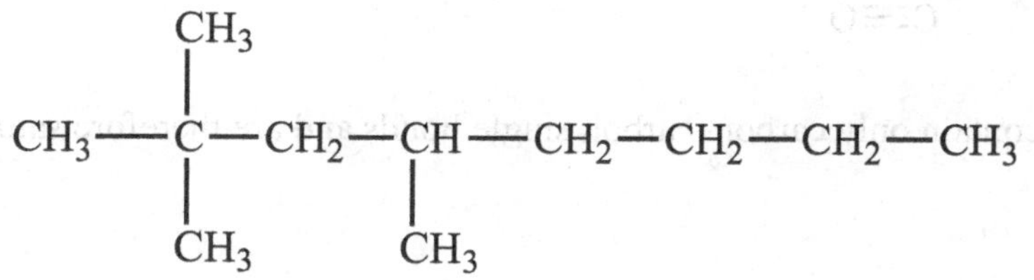

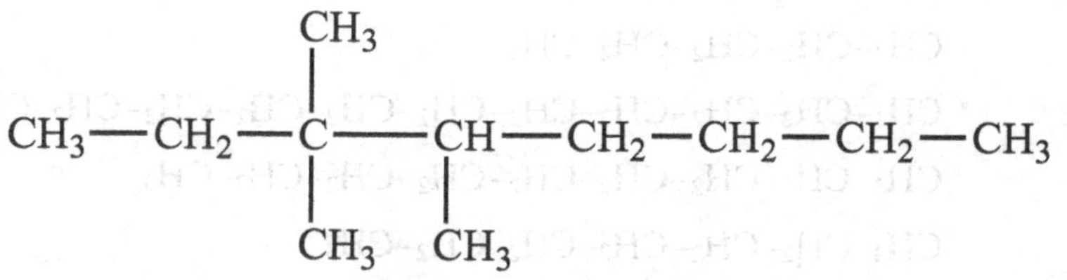

28. a.
$$CH_3-\underset{\underset{CH_3}{|}}{\overset{\overset{CH_3}{|}}{C}}-CH_2-\underset{\underset{CH_3}{|}}{CH}-CH_2-CH_2-CH_2-CH_3$$

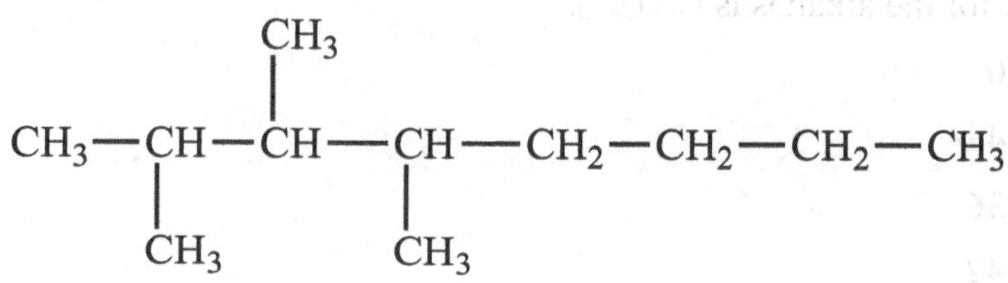

b.
$$CH_3-\underset{\underset{CH_3}{|}}{CH}-\overset{\overset{CH_3}{|}}{CH}-\underset{\underset{CH_3}{|}}{CH}-CH_2-CH_2-CH_2-CH_3$$

c.
$$CH_3-CH_2-\underset{\underset{CH_3}{|}}{\overset{\overset{CH_3}{|}}{C}}-\underset{\underset{CH_3}{|}}{CH}-CH_2-CH_2-CH_2-CH_3$$

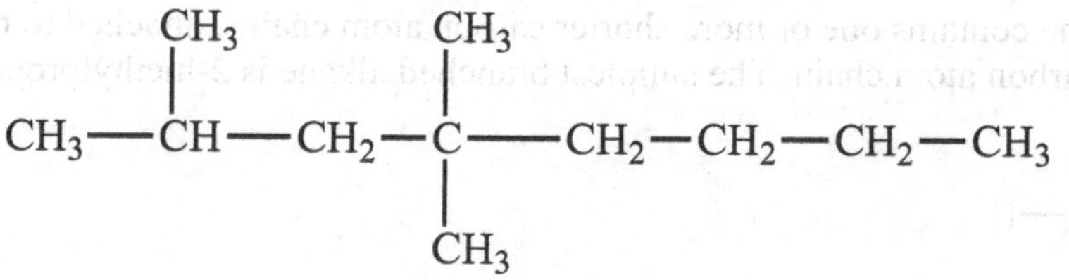

d.
$$CH_3-\underset{\underset{CH_3}{|}}{\overset{\overset{CH_3}{|}}{CH}}-CH_2-\underset{\underset{CH_3}{|}}{\overset{\overset{CH_3}{|}}{C}}-CH_2-CH_2-CH_2-CH_3$$

30. fractions

32. Tetraethyl lead was added to gasolines to prevent "knocking" of high efficiency automobile engines. The use of tetraethyl lead is being phased out because of the danger to the environment posed by the lead in this substance.

34. The combustion of alkanes has been used as a source of heat, light, and mechanical energy.
$$C_3H_8(g) + 5O_2(g) \rightarrow 3CO_2(g) + 4H_2O(g) + heat$$

Chapter 20: Organic Chemistry

36. dehydrogenation

38. a. $2C_8H_{18}(l) + 25O_2(g) \rightarrow 16CO_2(g) + 18H_2O(aq)$

b. $CH_3Cl(l) + Cl_2(g) \rightarrow CH_2Cl_2(l) + HCl(g)$

c. $CHCl_3(l) + Cl_2(g) \rightarrow CCl_4(l) + HCl(g)$

40. An alkyne is a hydrocarbon containing a carbon-carbon triple bond. The general formula is $C_nH_{2n-2}$.

42. The location of a double or triple bond in the longest chain of an alkene or alkyne is indicated by giving the *number* of the lowest number carbon atom involved in the double or triple bond.

44. Hydrogenation converts unsaturated compounds to saturated (or less unsaturated) compounds. In the case of a liquid vegetable oil, this is likely to convert the oil to a solid.

$$C_2H_4(g) + H_2(g) \rightarrow C_2H_6(g)$$

46. a. 5,5-dichloro-3,4-dimethyl-1-pentene

b. 4,5-dichloro-2-hexene (look for the *longest* chain)

c. 2,2,5-trimethyl-3-heptene

d. 5-methyl-1-hexyne

48. Shown are carbon skeletons:

C≡C—C—C—C—C     C—C≡C—C—C—C     C—C—C≡C—C—C

C≡C—C—C—C     C—C≡C—C—C     C≡C—C—C—C     C≡C—C—C
    |                 |                 |               |
    C                 C                 C               C
                                                        |
                                                        C

50. For benzene, a *set* of equivalent Lewis structures can be drawn, differing only in the *location* of the three double bonds in the ring. Experimentally, however, benzene does not demonstrate the chemical properties expected for molecules having *any* double bonds. We say that the "extra" electrons that would go into making the second bond of the three double bonds are delocalized around the entire benzene ring; this delocalization of the electrons explains benzene's unique properties.

52. When named as a substituent, the benzene ring is called the *phenyl* group. Two examples are:

$CH_2=CH-CH-CH_3$                $CH_3-CH-CH_2-CH_2-CH_2-CH_3$

3-phenyl-1-butene                    2-phenylhexane

Chapter 20:   Organic Chemistry

54. *ortho–* refers to adjacent substituents (1,2–); *meta–* refers to two substituents with one unsubstituted carbon atom between them (1,3–); *para–* refers to two substituents with two unsubstituted carbon atoms between them (1,4–).

56.  a.  3,4-dibromo-1-methylbenzene, 3,4-dibromotoluene
     b.  naphthalene
     c.  4-methylphenol; 4-hydroxytoluene
     d.  1,4-dinitrobenzene, *p*-dinitrobenzene

58.  a.  carboxylic (organic) acids
     b.  aldehydes
     c.  ketones
     d.  alcohols

60. Primary alcohols have *one* hydrocarbon fragment (alkyl group) bonded to the carbon atom where the –OH group is attached. Secondary alcohols have *two* such alkyl groups attached, and tertiary alcohols contain *three* such alkyl groups. Examples are:

   ethanol (primary)

   $CH_3—CH_2—OH$

   2-propanol (secondary)

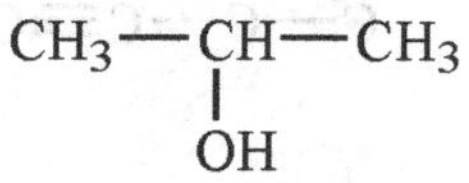

   $CH_3—CH—CH_3$
   $\phantom{CH_3—C}|$
   $\phantom{CH_3—}OH$

   2-methyl-2-propanol (tertiary)

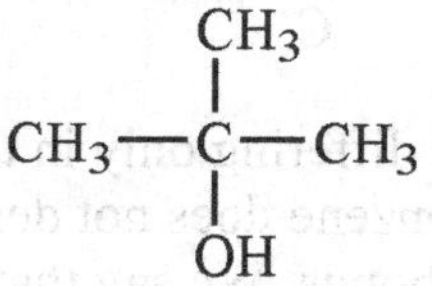

   $\phantom{CH_3—}CH_3$
   $\phantom{CH_3—C}|$
   $CH_3—C—CH_3$
   $\phantom{CH_3—C}|$
   $\phantom{CH_3—}OH$

62. 1-pentanol ($CH_3CH_2CH_2CH_2CH_2OH$) is a primary alcohol because the carbon atom where the –OH is attached contains only one *R* group.

64. The reaction is

   $C_6H_{12}O_6 \xrightarrow{yeast} 2CH_3–CH_2–OH + 2CO_2$

   The yeast necessary for the fermentation process are killed if the concentration of ethanol is over 13%. More concentrated ethanol solutions are most commonly made by distillation.

66. phenol; used to produce polymers for adhesives and plastics

68. Aldehydes and ketones both contain the carbonyl group C=O.

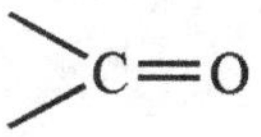

   $>C=O$

Chapter 20: Organic Chemistry

Aldehydes and ketones differ in the *location* of the carbonyl function: aldehydes contain the carbonyl group at the end of a hydrocarbon chain (the carbon atom of the carbonyl group is bonded only to at most one other carbon atom); the carbonyl group of ketones represents one of the interior carbon atoms of a chain (the carbon atom of the carbonyl group is bonded to two other carbon atoms).

70. The specific answers depend on your choice of alcohols. Here are representative reactions involving general primary and secondary alcohols:

$$R\text{--}CH_2\text{--}OH \xrightarrow{\text{mild oxidation}} R\text{--}CHO$$

$$R\text{--}CHOH\text{--}R' \xrightarrow{\text{mild oxidation}} R\text{--}C(=O)\text{--}R'$$

72. In addition to their systematic names (based on the hydrocarbon root, with the ending *–one*), ketones can also be named by naming the groups attached to either side of the carbonyl carbon as alkyl groups, followed by the word "ketone". Therefore, 2-butanone can also be named methyl ethyl ketone.

74. The structures are:

a. $CH_3\text{--}\overset{\overset{O}{\|}}{C}\text{--}CH_3$

b. $CH_3\text{--}\overset{\overset{O}{\|}}{C}\text{--}\overset{\overset{CH_3}{|}}{CH}\text{--}CH_3$

c. $CH_3\text{--}CH_2\text{--}C\overset{\nearrow O}{\underset{\searrow H}{}}$

d. $CH_3\text{--}\overset{\overset{CH_3}{|}}{\underset{\underset{CH_3}{|}}{C}}\text{--}\overset{\overset{}{\|}}{\underset{\underset{O}{}}{C}}\text{--}CH_2\text{--}CH_3$

76. Carboxylic acids are typically *weak* acids.

$CH_3\text{--}CH_2\text{--}COOH(aq) \rightleftharpoons H^+(aq) + CH_3\text{--}CH_2\text{--}COO^-(aq)$

78. 
a. $CH_3\text{--}CH_2\text{--}CH_2\text{--}CHO$

b. $CH_3\text{--}CH_2\text{--}COOH$

c. $CH_3\text{--}CH_2\text{--}CH_2\text{--}\overset{\overset{O}{\|}}{C}\text{--}O\text{--}CH_2\text{--}CH_2\text{--}CH_3$

80. The two main functional groups in acetylsalicylic acid (aspirin) are a carboxylic acid (–COOH) and an ester (RCOOR').

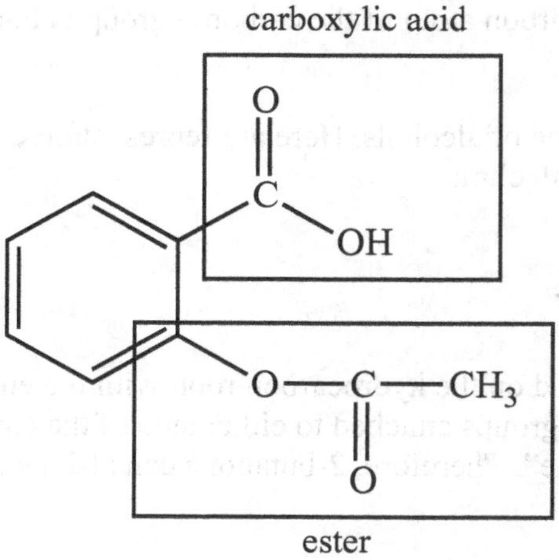


82. The structures are:

a. 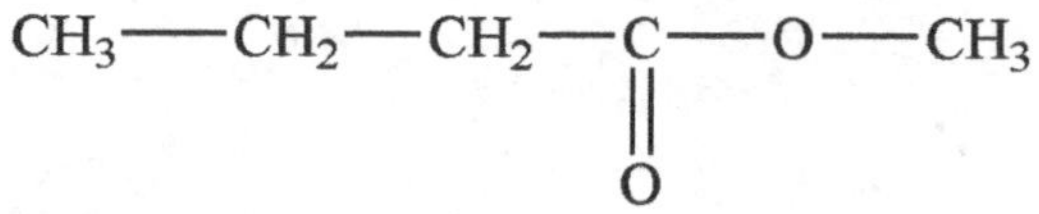

b. 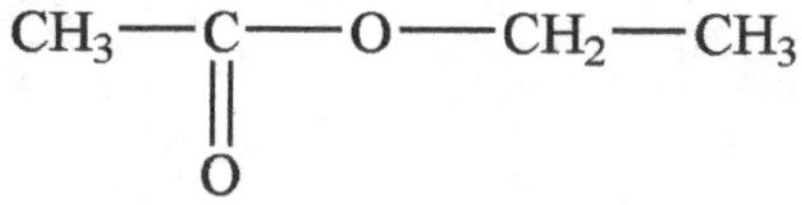

c. 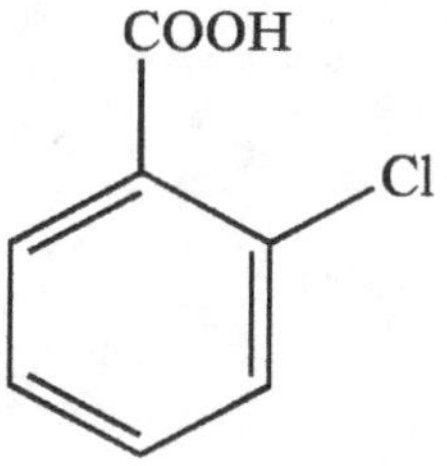

d. 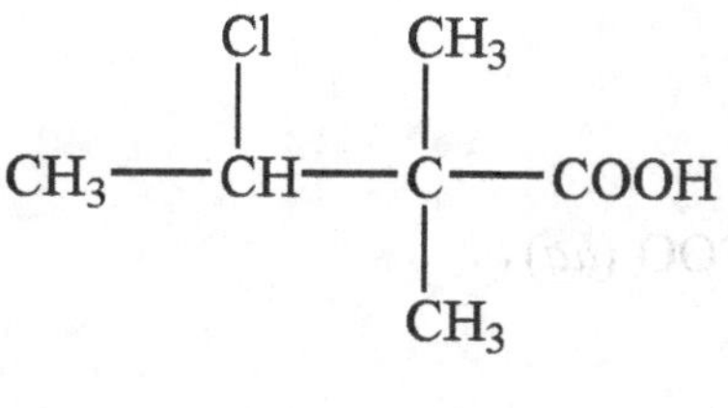

84. addition

86. Kevlar is a *co*-polymer since two different types of monomers combine to generate the polymer chain.

Chapter 20: Organic Chemistry

88. The structures are:

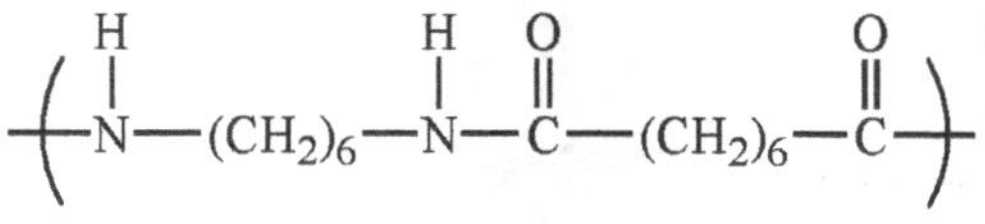

$$\left(-\underset{H}{N}-(CH_2)_6-\underset{H}{N}-\underset{\parallel}{\overset{O}{C}}-(CH_2)_6-\underset{\parallel}{\overset{O}{C}}-\right)$$

nylon

$$\left(-O-CH_2-CH_2-O-\underset{\parallel}{\overset{O}{C}}-\bigcirc-\underset{\parallel}{\overset{O}{C}}-\right)$$

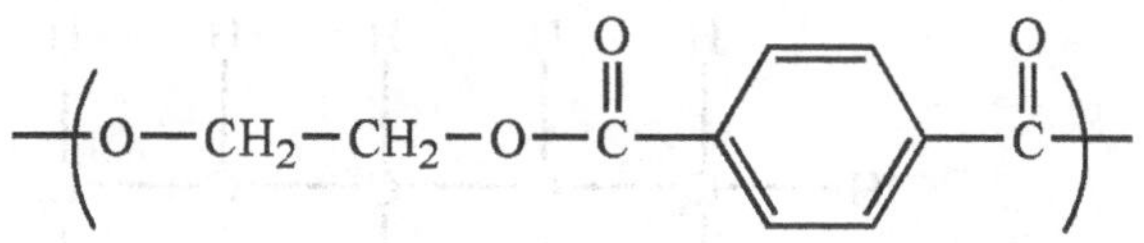

dacron

90. unsaturated

92. straight-chain or normal

94. *-ane*

96. number

98. anti-knocking

100. substitution

102. hydrogenation

104. functional

106. carbon monoxide

108. carbonyl

110. The correct answer is *d*. Organic molecules must contain carbon. The formula for magnesium sulfate is $MgSO_4$.

112. a.   ethane, $CH_3$–$CH_3$

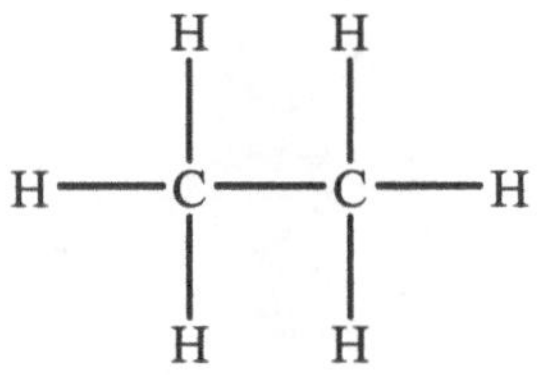

$$H-\underset{\underset{H}{|}}{\overset{\overset{H}{|}}{C}}-\underset{\underset{H}{|}}{\overset{\overset{H}{|}}{C}}-H$$

b.   butane, $CH_3$–$CH_2$–$CH_2$–$CH_3$

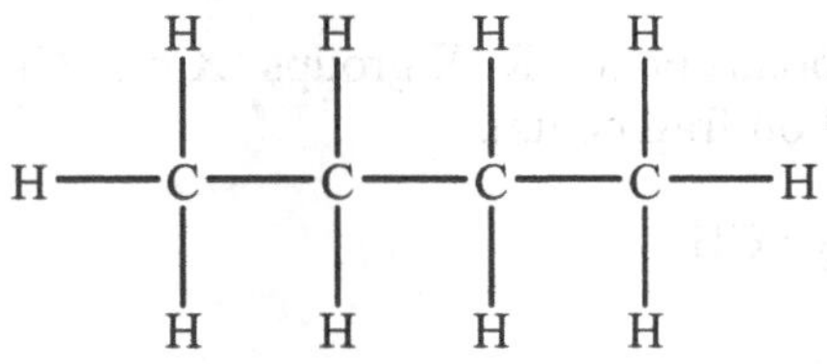

$$H-\underset{\underset{H}{|}}{\overset{\overset{H}{|}}{C}}-\underset{\underset{H}{|}}{\overset{\overset{H}{|}}{C}}-\underset{\underset{H}{|}}{\overset{\overset{H}{|}}{C}}-\underset{\underset{H}{|}}{\overset{\overset{H}{|}}{C}}-H$$

c. hexane, $CH_3-CH_2-CH_2-CH_2-CH_2-CH_3$

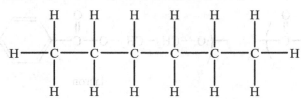

114. A saturated hydrocarbon is one in which all carbon-carbon bonds are single bonds, with each carbon atom forming bonds to four other atoms. The saturated hydrocarbons are called alkanes.

116. 
a. 2-butene
b. 3-methyl-1-butene
c. 1-butyne
d. 3-chloro-1-butene

118.
a. $CH_3-CH-CH-CH_2-CH_2-CH_2-CH_3$
       $\quad\ \ |\quad\ \ |$
       $\quad CH_3\ \ CH_3$

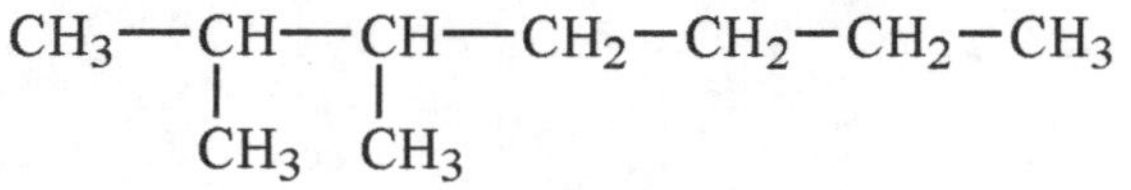

b. $\qquad\qquad\quad CH_3$
   $\qquad\qquad\quad |$
   $HO-CH_2-C-CH-CH_2-CH_2-CH_2-CH_2-CH_3$
   $\qquad\qquad\quad |\quad\ |$
   $\qquad\qquad\ CH_3\ \ Cl$

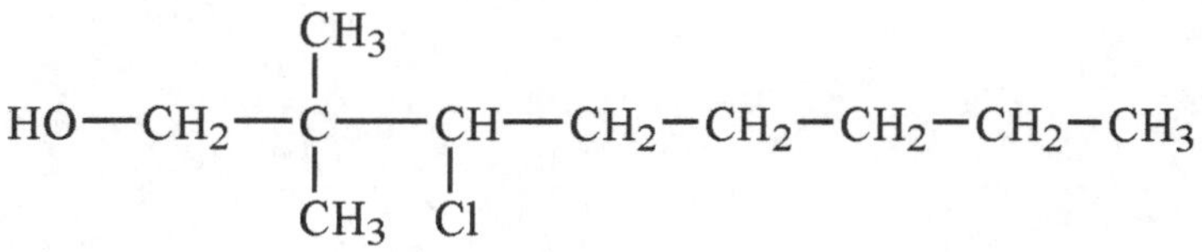

c. $CH_2=C-CH_2-CH_2-CH_2-CH_3$
   $\qquad\ \ |$
   $\qquad\ Cl$

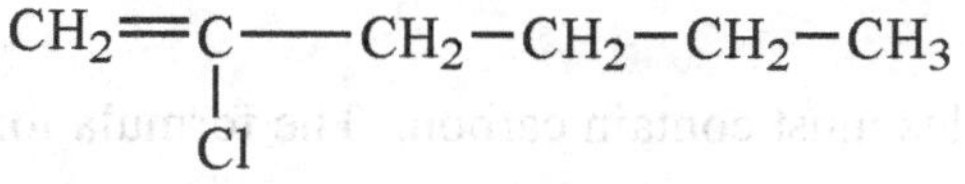

d. $Cl-CH_2-CH=CH-CH_2-CH_2-CH_3$

e. [structure: benzene ring with OH and CH_3 substituents in ortho positions]

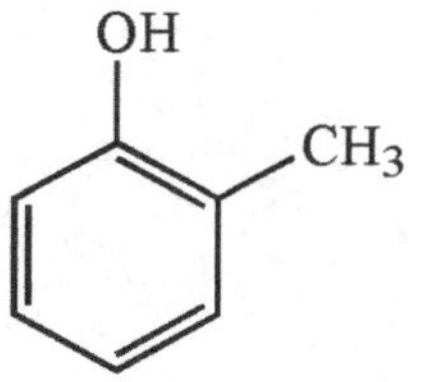

120. 1; Only isopropyl alcohol is a secondary alcohol.

122. The correct answer is *a* and *e*. Halohydrocarbons contain R–X groups (X = F, Cl, Br, I) and amines contain R–NH$_2$ groups (R = hydrocarbon fragments).

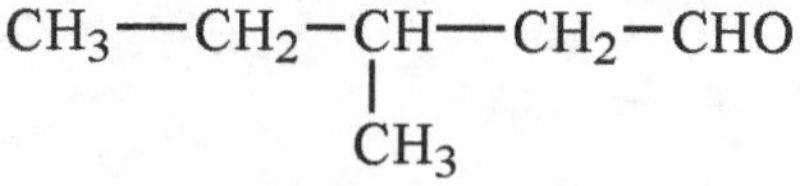

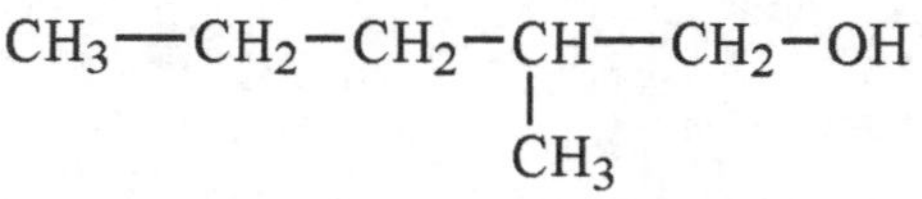

124.
a. $CH_3-C-CH_2-CH_2-CH_2-CH_2-CH_3$
   $\qquad\ \ ||$
   $\qquad\ \ O$

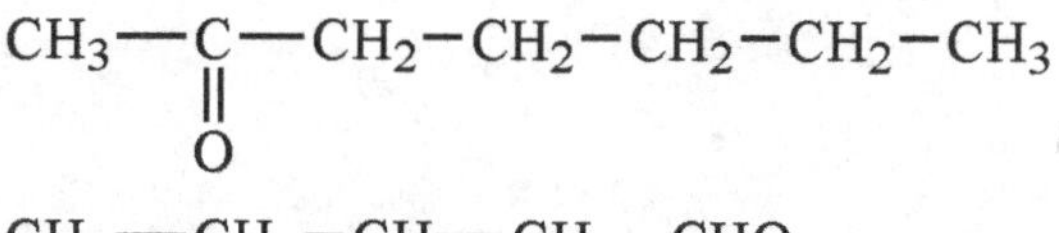

b. $CH_3-CH_2-CH-CH_2-CHO$
   $\qquad\qquad\ |$
   $\qquad\qquad CH_3$

c. $CH_3-CH_2-CH_2-CH-CH_2-OH$
   $\qquad\qquad\qquad\ |$
   $\qquad\qquad\qquad CH_3$

d. 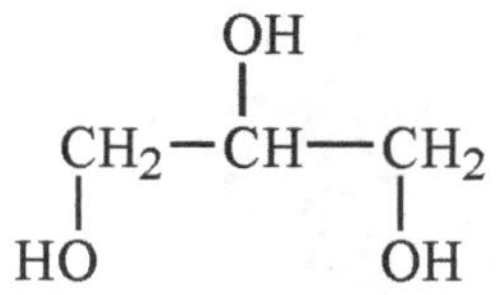
OH
CH2–CH–CH2
HO
OH

e. 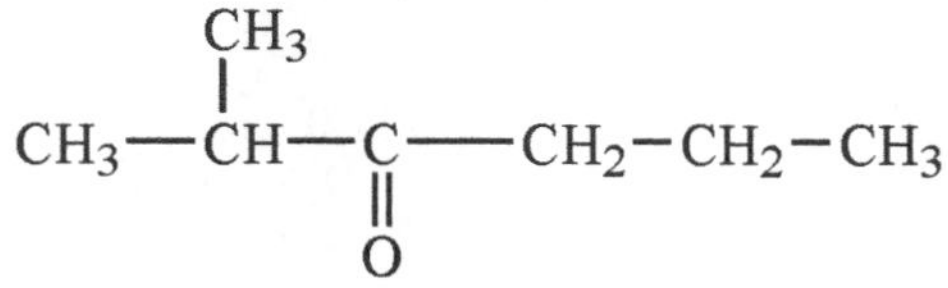
CH3
CH3–CH–C–CH2–CH2–CH3
O

126. 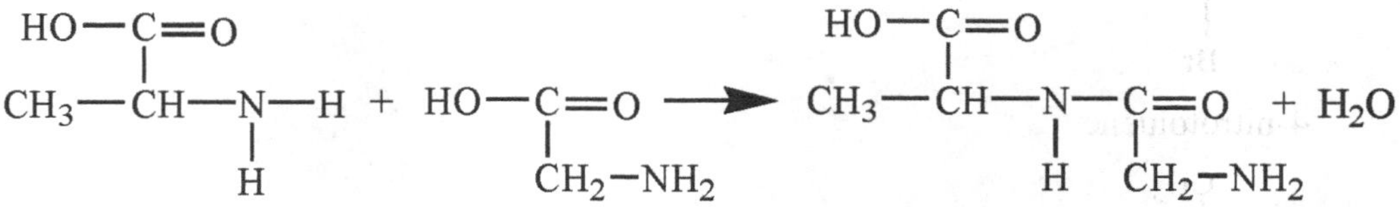
HO–C=O
CH3–CH–N–H + HO–C=O
H
CH2–NH2
HO–C=O
CH3–CH–N–C=O + H2O
H
CH2–NH2

128. a. 1,2-dichlorobenzene

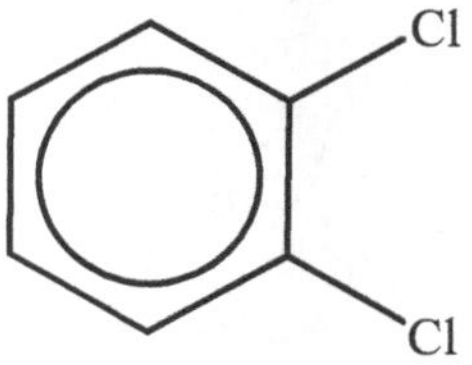
Cl
Cl

b. 1,3-dimethylbenzene

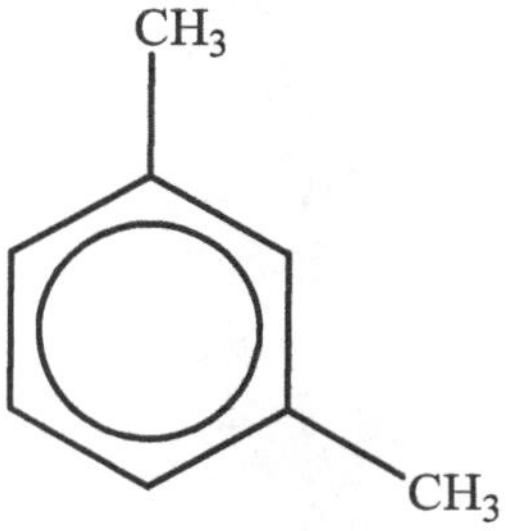
CH3
CH3

c. 3-nitrophenol

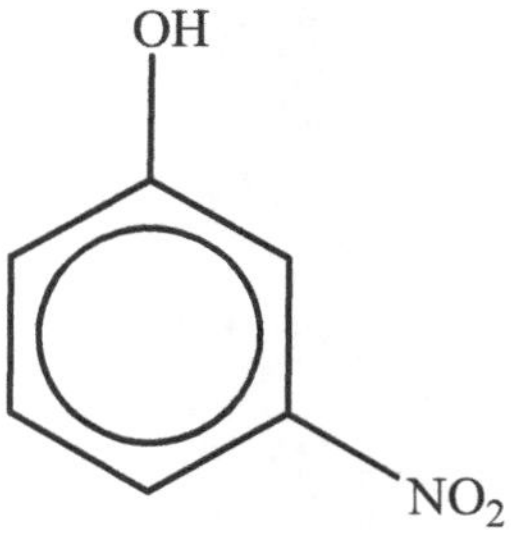
OH
NO2

d. *p*-dibromobenzene

e. 4-nitrotoluene

130. a. 2,3-dimethylbutane
 b. 3,3-diethylpentane
 c. 2,3,3-trimethylhexane
 d. 2,3,4,5,6-pentamethylheptane

132. a. $CH_3Cl(g)$
 b. $H_2(g)$
 c. $HCl(g)$

134. $CH{\equiv}C–CH_2–CH_2–CH_2–CH_2–CH_2–CH_3$  1-octyne
 $CH_3–C{\equiv}C–CH_2–CH_2–CH_2–CH_2–CH_3$  2-octyne
 $CH_3–CH_2–C{\equiv}C–CH_2–CH_2–CH_2–CH_3$  3-octyne
 $CH_3–CH_2–CH_2–C{\equiv}C–CH_2–CH_2–CH_3$  4-octyne

136. a. carboxylic acid
 b. ketone
 c. ester
 d. alcohol (phenol)

138. The correct answer is *e*. An ester has the general formula R-COO-R', and a carboxylic acid has the general formula R-COOH. A ketone has the general formula R-CO-R', and an aldehyde has the general formula R-COH.

140.  a. 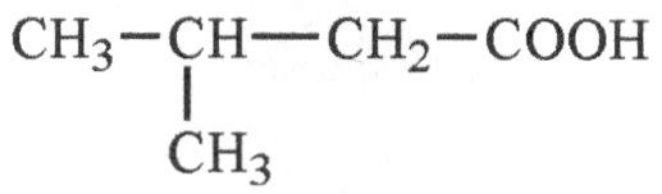  CH₃—CH—CH₂—COOH
              |
              CH₃

   b.  
   $$\text{benzoic acid with Cl substituent (2-chlorobenzoic acid)}$$

   (structure: benzene ring with -C(=O)-OH group and -Cl on adjacent carbon)

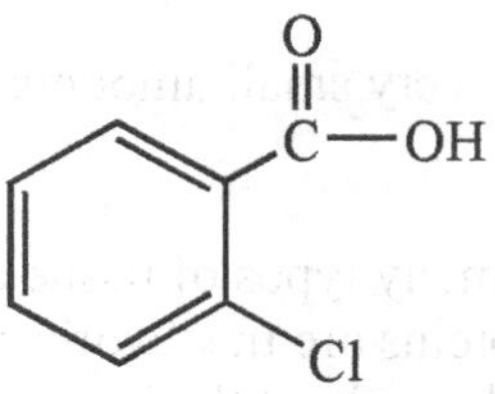

   c.  CH₃–CH₂–CH₂–CH₂–CH₂–COOH

   d.  CH₃–COOH

142.  a.  pentane
   b.  3-ethyl-2,5-dimethylhexane
   c.  4-ethyl-5-isopropyloctane

144.  a.  2-methyl-1-butene
   b.  2,4-dimethyl-1,4-pentadiene
   c.  6-ethyl-2-methyl-4-octene
   d.  3-bromo-1-heptyne
   e.  7-chloro-2,5,5-trimethyl-3-heptyne
   f.  4-ethyl-3-methyl-1-octyne

146.  2-chloropropanoic acid

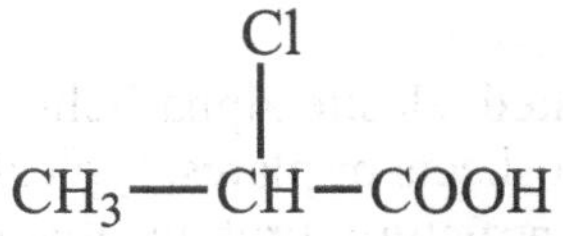

              Cl
              |
   CH₃—CH—COOH

# CHAPTER 21

# Biochemistry

2. Trace elements are those elements present in the body in only very small amounts, but which are essential to many biochemical processes in the body.

4. Fibrous proteins provide structural integrity and strength for many types of tissue and are the main components of muscle, hair, and cartilage. Globular proteins are the "worker" molecules of the body, performing such functions as transporting oxygen throughout the body, catalyzing many of the reactions in the body, fighting infections, and transporting electrons during the metabolism of nutrients.

6. cysteine and asparagine; A side chain is nonpolar if it is mostly hydrocarbon in nature (like alanine). Polar side chains may contain the hydroxyl group (–OH), the sulfhydryl group (–SH), or a second amino (–NH$_3$) or carboxyl (–COOH) group. Cysteine contains a sulfhydryl group and asparagine contains a second amino group.

8. The amino acid will be hydrophilic if the R group is polar, and hydrophobic if the R group is nonpolar. Serine is a good example of an amino acid in which the R group is polar. Leucine is a good example of an amino acid with a nonpolar R group.

10. There are six tripeptides possible.

    cys-ala-phe        ala-cys-phe        phe-ala-cys
    cys-phe-ala        ala-phe-cys        phe-cys-ala

12. The primary structure of a protein is the specific *sequence* of amino acids in the peptide chain. Adjacent amino acids are connected to each other by peptide (amide) linkages.

14. Long, thin, resilient proteins, such as hair, typically contain elongated, elastic alpha-helical protein molecules. Other proteins, such as silk, which in bulk form sheets or plates, typically contain protein molecules having the beta pleated sheet secondary structure. Proteins that do not have a structural function in the body, such as hemoglobin, typically have a globular structure.

16. pleated sheet

18. disulfide linkage

20. Oxygen is transported by the protein *hemoglobin*.

22. cytochromes

24. Amino acids contain both a weak-acid and a weak-base group, and thus they can neutralize both bases and acids, respectively.

26. The substrate and enzyme attach to each other in such a way that the part of the substrate where the reaction is to occur occupies the active site of the enzyme. After the reaction occurs, the

Chapter 21: Biochemistry

products are liberated, and the enzyme is ready for a new substrate. We can represent enzyme catalysis by the following steps:

Step 1: The enzyme E and the substrate S come together.

$$E + S \rightleftharpoons E \cdot S$$

Step 2: The reaction occurs to give the product P, which is released from the enzyme.

$$E \cdot S \rightarrow E + P$$

After the product is released, the enzyme is free to engage another substrate. Because this process occurs so rapidly, only a tiny amount of enzyme is required.

28. The lock-and-key model for enzymes indicates that the structures of an enzyme and its substrate must be *complementary*, so that the substrate can approach and attach itself along the length of the enzyme at the enzyme's active sites. A given enzyme is intended to act upon a particular substrate: the substrate attaches itself to the enzyme, is acted upon, and then moves away from the enzyme. If a different molecule has a similar structure to the substrate, this other molecule may also be capable of attaching itself to the enzyme. But since this molecule is not the enzyme's proper substrate, the enzyme may not be able to act upon the molecule, and the molecule may remain attached to the enzyme preventing proper substrate molecules from approaching the enzyme (irreversible inhibition). If the enzyme cannot act upon its proper substrate, then the enzyme is said to be inhibited. Irreversible inhibition might be a desirable feature in an antibiotic, which would bind to the enzymes of a bacteria and prevent the bacteria from reproducing, thereby preventing or curing an infection.

30. polymers

32. A hexose sugar is a carbohydrate containing 6 carbon atoms in the chain. A sketch of the straight-chain galactose is shown below.

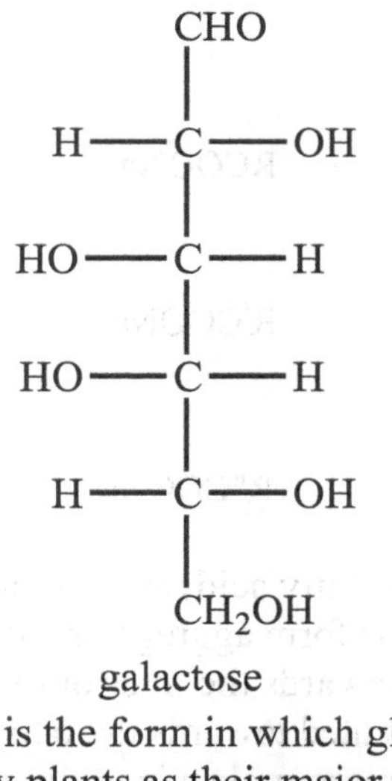

```
        CHO
         |
   H ─── C ─── OH
         |
  HO ─── C ─── H
         |
  HO ─── C ─── H
         |
   H ─── C ─── OH
         |
        CH₂OH
       galactose
```

34. Starch is the form in which glucose is stored by plants for later use as cellular fuel. Cellulose is used by plants as their major structural component. Although starch and cellulose are both polymers of glucose, the linkage between adjacent glucose units differs in the two polysaccharides. Humans do not possess the enzyme needed to hydrolyze the linkage in cellulose.

36. ribose (aldopentose); arabinose (aldopentose); ribulose (ketopentose); glucose (aldohexose); mannose (aldohexose); galactose (aldohexose); fructose (ketohexose).

## Chapter 21: Biochemistry

38. smaller

40. Uracil (RNA only); cytosine (DNA, RNA); thymine (DNA only); adenine (DNA, RNA); guanine (DNA, RNA)

42. An overall DNA molecule consists of two chains of nucleotides, with the organic bases on the nucleotides arranged in complementary pairs (cytosine with guanine, and adenine with thymine). The structures and properties of the organic bases are such that these pairs fit together well and allow the two chains of nucleotides to form the double helix structure. When DNA replicates, it is assumed the double helix unwinds, and then new molecules of the organic bases come in and pair up with their respective partner on the separated nucleotide chains, thereby replicating the original structure. See Figure 21.20

44. Lipids are a group of substances defined in terms of their solubility characteristics: lipids are typically oily, greasy substances that are not very soluble in water.

46. A triglyceride typically consists of a glycerol backbone, to which three separate fatty acid molecules are attached by ester linkages.

$$R-\overset{O}{\underset{\|}{C}}-O-CH_2$$
$$\phantom{R-C-O-}CH-O-\overset{O}{\underset{\|}{C}}-R'$$
$$R''-\overset{O}{\underset{\|}{C}}-O-CH_2$$

48. "Soaps" are the salts of long-chain organic acids ("fatty acids"), most commonly either the sodium or potassium salt. Soaps are prepared by treating a fat or oil (a triglyceride) with a strong base such as NaOH or KOH. This breaks the ester linkages in the triglyceride, releasing three fatty acid anions and glycerol.

$$\begin{array}{c}CH_2-O-\overset{O}{\underset{\|}{C}}-R\\CH-O-\overset{O}{\underset{\|}{C}}-R'\\CH_2-O-\overset{O}{\underset{\|}{C}}-R''\end{array} + 3NaOH \longrightarrow \begin{array}{c}CH_2-OH\\CH-OH\\CH_2-OH\end{array} + \begin{array}{c}RCOONa\\R'COONa\\R''COONa\end{array}$$

50. Soaps have both a nonpolar nature (due to the long chain of the fatty acid) and an ionic nature (due to the charge on the carboxyl group). In water, soap anions form aggregates called micelles, in which the water-repelling hydrocarbon chains are oriented towards the interior of the aggregate, with the ionic, water-attracting carboxyl groups oriented towards the outside. Most dirt has a greasy nature. A soap micelle is able to interact with a grease molecule, pulling the grease molecule into the hydrocarbon interior of the micelle. When the clothing is rinsed, the micelle containing the grease is washed away. See Figures 21.22 and 21.23.

52. adrenocorticoid hormones; cortisol

Chapter 21:   Biochemistry

54. The bile acids are synthesized from cholesterol in the liver and are stored in the gall bladder. Bile acids such as cholic acid act as emulsifying agents for lipids and aid in their digestion.

56. h

58. l

60. t

62. e

64. f

66. q

68. o

70. n

72. x

74. c

76. a

78. The RNA molar mass range is much smaller than the molar mass of DNA. The molar mass of DNA depends on the complexity of the species, but human DNA may have a molar mass as large as 2 billion g/mol.

80. ester

82. thymine, guanine

84. transfer, messenger

86. cys-ala-phe; cys-phe-ala; phe-ala-cys; phe-cys-ala; ala-cys-phe; ala-phe-cys

88. unsaturated, saturated

90. ionic, nonpolar

92. steroids

94. The structures of steroids have a characteristic carbon ring structure of the type

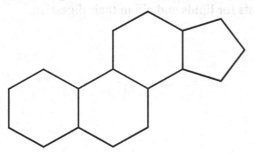

96. The polypeptide chain forms a coil or spiral. Such proteins are found in wool, hair, and tendons.

98. tendons, bone (with mineral constituents), skin, cartilage, hair, fingernails.

100. Collagen consists of three protein chains (each with α-helical structure) twisted together to form a superhelix. The result is a long, relatively narrow protein. Collagen functions as the raw material from which tendons are constructed.

102. pentoses (5 carbons); hexoses (6 carbons); trioses (3 carbons)

104. In a strand of DNA, the phosphate group and the sugar molecule of adjacent nucleotides become bound to each other. The chain-portion of the DNA molecule, therefore, consists of alternating phosphate groups and sugar molecules. The nitrogen bases are found protruding from the side of this phosphate-sugar chain, bonded to the sugar molecules.

106. Phospholipids are esters of glycerol. Two fatty acids are bonded to the –OH groups of the glycerol backbone, with the third –OH group of glycerol bonded to a phosphate group. Having the two fatty acids, but also the polar phosphate group, makes the phospholipid lecithin a good emulsifying agent.